Fanny Masson

Molecular Dynamics Simulations of a Lesion in DNA

Fanny Masson

Molecular Dynamics Simulations of a Lesion in DNA

Autocatalytic or Enzymatic Repair and Formation of Thymine Dimer

Südwestdeutscher Verlag für Hochschulschriften

Impressum/Imprint (nur für Deutschland/ only for Germany)
Bibliografische Information der Deutschen Nationalbibliothek: Die Deutsche Nationalbibliothek verzeichnet diese Publikation in der Deutschen Nationalbibliografie; detaillierte bibliografische Daten sind im Internet über http://dnb.d-nb.de abrufbar.
Alle in diesem Buch genannten Marken und Produktnamen unterliegen warenzeichen-, marken- oder patentrechtlichem Schutz bzw. sind Warenzeichen oder eingetragene Warenzeichen der jeweiligen Inhaber. Die Wiedergabe von Marken, Produktnamen, Gebrauchsnamen, Handelsnamen, Warenbezeichnungen u.s.w. in diesem Werk berechtigt auch ohne besondere Kennzeichnung nicht zu der Annahme, dass solche Namen im Sinne der Warenzeichen- und Markenschutzgesetzgebung als frei zu betrachten wären und daher von jedermann benutzt werden dürften.

Verlag: Südwestdeutscher Verlag für Hochschulschriften Aktiengesellschaft & Co. KG
Dudweiler Landstr. 99, 66123 Saarbrücken, Deutschland
Telefon +49 681 37 20 271-1, Telefax +49 681 37 20 271-0, Email: info@svh-verlag.de
Zugl.: Zurich, University of Zurich, Diss., 2007

Herstellung in Deutschland:
Schaltungsdienst Lange o.H.G., Berlin
Books on Demand GmbH, Norderstedt
Reha GmbH, Saarbrücken
Amazon Distribution GmbH, Leipzig
ISBN: 978-3-8381-0513-0

Imprint (only for USA, GB)
Bibliographic information published by the Deutsche Nationalbibliothek: The Deutsche Nationalbibliothek lists this publication in the Deutsche Nationalbibliografie; detailed bibliographic data are available in the Internet at http://dnb.d-nb.de.
Any brand names and product names mentioned in this book are subject to trademark, brand or patent protection and are trademarks or registered trademarks of their respective holders. The use of brand names, product names, common names, trade names, product descriptions etc. even without a particular marking in this works is in no way to be construed to mean that such names may be regarded as unrestricted in respect of trademark and brand protection legislation and could thus be used by anyone.

Publisher:
Südwestdeutscher Verlag für Hochschulschriften Aktiengesellschaft & Co. KG
Dudweiler Landstr. 99, 66123 Saarbrücken, Germany
Phone +49 681 37 20 271-1, Fax +49 681 37 20 271-0, Email: info@svh-verlag.de

Copyright © 2009 by the author and Südwestdeutscher Verlag für Hochschulschriften Aktiengesellschaft & Co. KG and licensors
All rights reserved. Saarbrücken 2009

Printed in the U.S.A.
Printed in the U.K. by (see last page)
ISBN: 978-3-8381-0513-0

To my Grandmother, Φανή

Contents

Kurzfassung	ix
Abstract	xi
Résumé	xiii
Publications	xv

1 Introduction   1
     1.1 Pyrimidine Dimer 4
         1.1.1 Pyrimidine Dimer Repair 5
         1.1.2 Pyrimidine Dimer Formation 6

2 Computational Methods   9
     2.1 Classical Molecular Dynamics 11
     2.2 *Ab initio* Molecular Dynamics 13
         2.2.1 Density Functional Theory 13
         2.2.2 Born-Oppenheimer Molecular Dynamics 16
         2.2.3 Car-Parrinello Molecular Dynamics 17
     2.3 Hybrid QM/MM Molecular Dynamics 18
     2.4 Metadynamics 19
     2.5 Excited-State Calculations 20

3 Thymine Dimer Repair and Formation in vacuo   25
     3.1 Introduction 27
     3.2 Methods 29
         3.2.1 Computational Details 29
     3.3 Results and Discussion 30

	3.4	Conclusion	38
4	**Self-Repair of Thymine Dimer**		**41**
	4.1	Abstract	42
	4.2	Introduction	43
	4.3	Methods	45
		4.3.1 Structural Model	45
		4.3.2 Classical MD Simulation	46
		4.3.3 QM/MM MD Simulation	46
	4.4	Results and Discussion	49
	4.5	Bridging the time scale: Free Energy simulations	55
	4.6	Conclusion	57
	4.7	Appendix	59
5	**Thymine Dimer Repair by DNA Photolyase**		**63**
	5.1	Introduction	65
	5.2	Methods	68
		5.2.1 Structural Model	68
		5.2.2 Classical MD Simulation	69
		5.2.3 QM/MM MD Simulation	69
	5.3	Results and Discussion	71
	5.4	Conclusion	83
	5.5	Appendix	84
6	**Metadynamics Study of Thymine Dimer Formation in DNA**		**95**
	6.1	Introduction	97
	6.2	Methods	99
		6.2.1 Structural Model	99
		6.2.2 QM/MM MD Simulations	99
		6.2.3 Metadynamics	101
	6.3	Results and Discussion	102
		6.3.1 Thymine Dimer Formation on Ground-State FES	102
		6.3.2 Vertical Excitation Energies of the Lowest Triplet- and Singlet-States	104
		6.3.3 Thymine Dimer Formation on Excited-State FES	107

CONTENTS

6.4 Conclusion 116

Outlook 119

Acknowledgments 121

Curriculum Vitae 123

Kurzfassung

Die Bildung des Thymin-Dimers ist die am häufigsten auftretende Schädigung innerhalb der DNA, verursacht durch ultraviolettes Licht. Es bildet sich zwischen zwei Nebenthyminen durch eine [2 + 2] Photocycloaddition. Störungen der DNA-Funktionalität können komplexe biologische Prozesse wie Apoptose, eine Beeinträchtigung des Immunsystems und Krebs hervorrufen. Die Reparatur- und Bildungs-Reaktionen des Thymin-Dimers wurden computergestützt untersucht. Ziel dieser Dissertation ist das Sichtbarmachen der Schritte des Mechanismus dieser Prozesse. Diese sind bei experimentellen Techniken nicht direkt zugänglich.

Als erster Schritt zur Modelierung der Reparaturreaktion dieser Störung in der DNA wurden GGA/DFT-Rechnungen der elektronischen Struktur in der Gasphase durchgeführt. Während der Elektronenaufnahme von dem Dimer geschieht ein spontaner Bruch der C5-C5' Bindung. Metadynamik-Simulationen zeigen eine Aktivierungsenergie des späteren Bruchs der C6-C6' Bindung von 6 kcal/mol. Rechnungen von vertikalen Anregungsenergien wurden ebenso ausgeführt, um zusätzliche Einsichten in die Bildungsreaktion des Dimers zu erhalten.

Wir haben einen hybriden Quantum/Klassischen (QM/MM) Ansatz benutzt, um die gesamte Umgebung der Störung in unseren Modellen aufzunehmen. Wir haben zuerst die Selbst-Reparatur des Thymin-Dimers in der DNA behandelt. Eine Menge von 7 statistisch repräsentativen QM/MM Moleküldynamik-Trajektorien wurde analysiert. Unsere Rechnungen bestätigten die experimentellen Ergebnisse für einen Eigen-Reparatur-Mechanismus, bei dem wir einen asynchron konzertierten Spaltungs-Mechanismus vorhersagen. Der Bruch der C5-C5' Bindung verläuft ohne Barriere, wogegen beim Bruch der C6-C6' Bindung eine kleine Barriere von freien Reaktionsenthalpie auftritt. Bei Verwendung von Metadynamik wurde eine obere Schranke von 2.5 kcal/mol für diese Barriere eingeführt.

Die theoretischen Untersuchungen bestätigten einerseits die thermodynamische Durchführbarkeit und andererseits die kinetische Durchführbarkeit des Eigen-Reparatur Prozesses.

Wir haben auch die Reparatur des Thymin-Dimers in dem aktiven Zentrum von DNA Photolyase studiert. DNA Photolyase ist ein sehr effizientes, von Licht angetriebenes Enzym, das die Störung direkt repariert, indem es ein Elektron von dem Flavin-Kofaktor transferiert. In Analogie zu der Eigen-Reparatur-Reaktion wurde herausgefunden, dass der Spaltungs-Mechanismus des Cyclobutan-Rings asynchron konzertiert ausgeführt wird. Ausserdem wurden Eigenschaften, die den gesamten Spaltungs-Mechanismus charakterisieren von unseren Simulationen aufgezeigt: eine durchgehende Umorientierung des Lösungsmittels im aktiven Zentrum, ein Proton-Transfer von Glu283 zum Thymin-Dimer, sowie starke Wechselwirkungen zwischen kationischen Seitenketten von Arg232 und Arg350 und dem Dimer. Unsere Resultate verdeutlichen die wichtige Rolle von Wasserstoffbrücken im aktiven Zentrum bei der Stabilisierung des Thymin-Dimer Anions. Dies führt zu hohen Reparatur-Quantenausbeuten.

Zuletzt werden die Singlet- und Triplet-Reaktionswege bei der Bildung des Thymin-Dimers in einem DNA Dekamer unter Anwendung von QM/MM Metadynamik-Simulationen erforscht. Vorstufen des Thymin-Dimers konnten auf beiden Oberflächen identifiziert werden. Diese Konformationen sind nicht direkt von der Grundzustand-Oberfläche zu erreichen. Dafür muss eine hohe Barriere auf den Oberflächen der angeregten Zustände überwunden werden. Gleichwohl bieten Moleküldynamik-Simulationen neue Einsichten in die Relaxierungswege der angeregten Zustände, die zu dem Dimer führen. Es wurde gezeigt, dass die Triplet-Reaktion über ein Diradikal-Zwischenprodukt verläuft, das zum Thymin-Dimer via eines Kreuzungspunktes mit S_0 zerfällt. Im Gegensatz dazu ist der Singlet-Mechanismus nicht mit einem stabilen Zwischenprodukt entlang des Weges in Richtung des Kreuzungspunktes mit S_0 verknüpft.

Abstract

Thymine dimer is the most abundant lesion in ultraviolet (UV)-irradiated DNA and is formed betweeen two adjacent thymine nucleobases via a [2 + 2] photocycloaddition. By disrupting the function of DNA, this lesion can trigger complex biological responses, including apoptosis, immune suppression, and carcinogenesis . The thymine dimer repair and formation reactions have been computationally investigated. The goal of this thesis is to elucidate key steps in the mechanism of these processes that are not readily accessible by experimental techniques.

As a first step toward modeling the repair reaction of this lesion in DNA, electronic structure calculations have been carried out in the gas phase at the GGA/DFT level. Upon electron uptake by the dimer, a spontaneous cleavage of the C5-C5' bond occurs. According to metadynamics simulations, the activation energy of subsequent C6-C6' bond breaking amounts to 6 kcal/mol. Calculations of vertical excitation energies were performed as well to get a first insight into the formation of the dimer.

In order to include the full environment of the lesion in our models, we used a hybrid quantum/classical (QM/MM) molecular dynamics approach. We first dealt with self-repair of thymine dimer in DNA. A set of 7 statistically representative QM/MM molecular dynamics trajectories was analyzed. Our calculations confirmed the experimental results of a self-repair mechanism, predicting an asynchronously concerted splitting mechanism in which C5-C5' bond breaking is barrierless while C6-C6' bond breaking is characterized by a small free energy barrier. Using metadynamics, an upper bound of 2.5 kcal/mol for this barrier was estimated. The theoretical investigations confirmed both the thermodynamical and kinetic feasibility of the self-repair process.

We studied the repair of the thymine dimer in the active site of DNA photolyase as well. DNA photolyase is a highly efficient light-driven enzyme which directly repairs the lesion by transferring an electron from its flavin cofactor. In analogy to the self-repair reaction, we find that the splitting mechanism of the cyclobutane ring is asynchronously concerted. Moreover, key features characterizing the overall splitting mechanism have been disclosed by our simulations: a continuous solvation reordering of the active site, a proton transfer from Glu283 to the thymine dimer and tight interactions between cationic side chains of Arg232 and Arg350 and the dimer. This suggests the important role of the active-site hydrogen-bond pattern in stabilizing the thymine dimer anion, leading to high repair quantum yields.

Finally, we explored the singlet and triplet reaction pathways of the thymine dimer formation in a DNA decamer by means of QM/MM metadynamics simulations. Precursors of the thymine dimer could be identified on both surfaces, but these conformations are not directly accessible from the ground-state surface and a significant barrier must be overcome on the excited state surfaces to reach them. Nonetheless, molecular dynamics simulations yield new insights into the relaxation pathways in the excited states leading to the thymine dimer. It is found that the triplet reaction proceeds over a diradical intermediate which decays to the thymine dimer via a crossing point with S_0. In contrast, the singlet mechanism does not involve any stable intermediate along the path towards a point of intersection with S_0.

Résumé

La formation du dimère de thymine est le dommage le plus fréquemment infligé a l'ADN par la lumière ultraviolette. Il se forme entre deux thymines adjacentes via une photocycloaddition [2 + 2]. En inhibant le mécanisme de réplication de l'ADN, cette lésion peut induire des réponses biologiques complexes, comme l'apoptose, l'immuno-suppression et la carcinogénèse. Les réactions de réparation et de formation du dimère de thymine ont été explorées computationnellement. Le but de cette thèse est d'élucider des étapes clé dans le mécanisme de ces processus qui ne sont pas accessibles directement par des techniques expérimentales.

Comme première étape vers la modélisation de la réaction de réparation de cette lésion dans l'ADN, des calculs de la structure électronique ont été menés en phase gazeuse au niveau GGA/DFT. Lors de la capture d'un électron par le dimère, une rupture spontanée de la liaison C5-C5' se produit. Selon des simulations de métadynamique, l'énergie d'activation de la rupture ultérieure de la liaison C6-C6' se monte à 6 kcal/mol. Des calculs des énergies d'excitation verticale ont également été effectués afin d'avoir un premier aperçu de la formation du dimère.

Afin d'inclure l'environnement complet de la lésion dans nos modèles, nous avons utilisé une approche hybride quantique/classique (QM/MM) de dynamique moléculaire. Nous avons d'abord traité l'auto-réparation du dimère de thymine dans l'ADN. Un ensemble de 7 trajectoires de dynamique moléculaire QM/MM statistiquement représentatives ont été analysées. Nos calculs confirment les résultats expérimentaux d'un mécanisme d'auto-réparation, en prédisant un mécanisme de scission asynchroniquement concerté, dans lequel la rupture de la liaison C5-C5' est dépourvue de barrière alors que la rupture de la liaison C6-C6' est caractérisée par une petite barrière d'énergie libre. En utilisant la métadynamique, une limite supérieure de 2.5 kcal/mol pour cette barrière a été estimée. Les investigations théoriques ont confirmé d'une part la faisabilité thermodynamique et d'autre part la faisabilité cinétique du processus d'auto-réparation.

Nous avons également étudié la réparation du dimère de thymine dans le site actif de la photolyase ADN. La photolyase ADN est une enzyme très efficace qui tire son énergie de la lumière pour réparer directement la lésion en transférant un électron de son cofacteur flavine. De manière analogue à la réaction d'auto-réparation, nous démontrons que le mécanisme de scission du cycle cyclobutane est asynchroniquement concerté. De plus, des éléments clé caractérisant le mécanisme global de scission ont été révélés par nos simulations : une redistribution continue des molécules de solvent dans le site actif, un transfert de proton de Glu283 au dimère de thymine et des interactions fortes entre les chaînes latérales cationiques de Arg232 et Arg350 et le dimère. Ces résultats soulignent l'importance du réseau de liaisons-hydrogène du site actif dans la stabilisation de l'anion du dimère de thymine, contribuant ainsi à des rendements quantiques de réparation élevés.

Finalement, nous avons exploré les chemins de réaction singulet et triplet de formation du dimère de thymine dans un décamère d'ADN par le biais de simulations de métadynamique QM/MM. Des précurseurs du dimère de thymine ont pu être identifiés sur les deux surfaces, mais ces conformations ne sont pas accessibles directement à partir de la surface de l'état fondamental et une barrière significative doit être franchie sur les surfaces d'états excités pour les atteindre. Néanmoins, les simulations de dynamique moléculaire apportent une nouvelle compréhension des chemins de relaxation dans les états excités menant au dimère de thymine. Il est démontré que la réaction dans l'état triplet passe par un intermédiaire diradical qui mène au dimère de thymine via un point d'intersection avec S_0. Par opposition, le mécanisme singulet n'implique pas d'intermédiaire stable le long du chemin vers un point d'intersection avec S_0.

Publications

Chapter 4

Fanny Masson, Teodoro Laino, Ivano Tavernelli, Ursula Rothlisberger and Jürg Hutter,
"Computational Evidence for Self-Repair of Thymine Dimer in Duplex DNA",
Submitted to *Journal of the American Chemical Society* (2007).

Chapter 5

Fanny Masson, Teodoro Laino, Ursula Rothlisberger and Jürg Hutter,
"A QM/MM Investigation of Thymine Dimer Repair by DNA Photolyase",
To be submitted to *Biochemistry* (2007).

Chapter 6

Fanny Masson, Ivano Tavernelli, Ursula Rothlisberger and Jürg Hutter,
"A Mixed QM/MM Metadynamics Study of Thymine Dimer Formation in DNA",
To be submitted.

Related publication

Denis Bucher, Fanny Masson, J. Samuel Arey, Jürg Hutter and Ursula Rothlisberger,
"DNA Repair Enzymes via Hybrid QM/MM simulations",
Manuscript in preparation.

Chapter 1

Introduction

In this thesis, we use techniques of computational chemistry to explore the repair and the formation mechanisms of the predominant UV-induced lesion in DNA, namely the thymine dimer. By disrupting the function of DNA, this lesion can trigger complex biological responses, including apoptosis, immune suppression, and carcinogenesis [1]. Here, we will first provide some insight into DNA damage and give a short overview of the field of pyrimidine dimer research.

As organisms reproduce, they copy their DNA. The copying is not always exact; occasionally mistakes are made. These may occur as random errors in copying, or they may be results of damage the DNA has suffered from radiation or chemical mutagens. In any event, these alterations will appear as mutations in the DNA of the next and subsequent generations.

For each polypeptide chain an organism produces, there exists a corresponding gene. The nucleotide sequence in that gene dictates, via the genetic code, the amino acid sequence of the protein. The effects of mutations on the functionality of the protein product, and therefore on the organism itself, can be quite varied. For example, base substitutions may, in some cases, be neutral in effect, either not changing the amino acid coded for or changing it to another that functions equally well at that position in that protein. More often, the result is deleterious. Occasionally, such mutations increase the efficiency of a protein, and the mutated organims may be selected for future generations. By contrast, nonsense mutations, which introduce a stop signal that results in a premature release of the polypeptide chain, almost always produce inactive protein products. If the protein is important to the life of the organism, such mutations are strongly selected against in the course of evolution.

The ultraviolet (UV) component of sunlight is a ubiquitous DNA-damaging agent. Its major effect is to covalently link adjacent pyrimidine residues along a DNA strand, resulting in formation of cyclobutane pyrimidine dimers and pyrimidine-pyrimidone (6-4) photoproducts (Figure 1.1). Note that this is not a rare event: every second we are in the sun, 50 to 100 of these dimers are formed in each skin cell ! [1] Such pyrimidine dimers cannot fit into a double helix, and so replication and gene expression are blocked until the lesion is removed.

---

[1]http://www.rcsb.org/pdb/static.do?p=education_discussion/molecule_of_the_month/pdb91_1.html

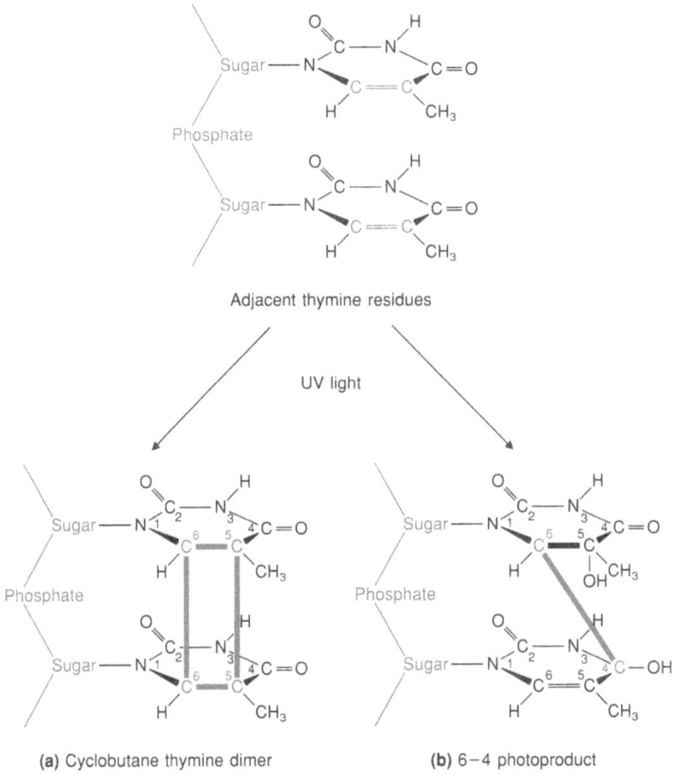

Introduction

Figure 1.1: (a) Structure of a cyclobutane pyrimidine dimer. Ultraviolet light stimulates the formation of a four-membered cyclobutyl ring (green) between two adjacent pyrimidines on the same DNA strand by acting on the 5,6 double bonds. (b) Structure of the 6-4 photo-product. The structure forms most prevalently with 5-C-C-3 and 5-T-C-3, between the C-6 and C-4 positions of two adjacent pyrimidines, causing a significant perturbation in local structure of the double helix. (Figure adapted from E. C. Friedberg, DNA Repair.)

To protect the genetic message, a wide range of DNA-repair enzymatic systems are present in most organisms [2]. Of the half-dozen known DNA repair processes, most involve removal of the damaged nucleotides, followed by replacement of the excised region using information encoded in the complementary (undamaged) strand. In contrast, the photochemical cleavage of pyrimidine dimers by the enzyme DNA photolyase is a reaction that directly changes the damaged bases, rather than removing them, and as such is a typical example of direct repair [3] (Figure 1.2). Note that placental mammals, including humans, do not have DNA photolyase and employ nucleotide excision repair (NER) pathways to remove pyrimidine dimers.

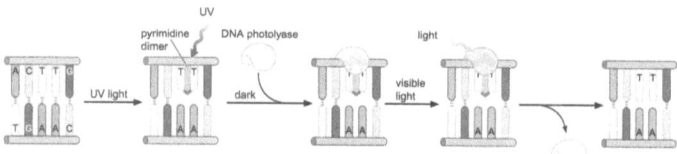

Figure 1.2: Repair of a UV-induced pyrimidine photodimer by a photoreactivating enzyme, or photolyase. The enzyme recognizes the photodimer and binds to it. When light is present, the photolyase uses its energy to split the dimer into the original monomers. (Figure adapted from J. D. Watson, Molecular Biology of the Gene, 3d ed.)

1.1 Pyrimidine Dimer

The first chemical evidence for the photoinduced formation of a pyrimidine dimer ($Pyr <> Pyr$) was obtained in the early 1960s by exposing thymine in frozen aqueous solutions to far-UV light [4]. The next major discovery dealt with the characterization of $Pyr <> Pyr$ as a major far-UV lesion within DNA [5]. This finding gave a strong impetus to the development of the photochemistry of nucleic acids, as illustrated in the following decade by the report of numerous studies related to the formation, isolation, and structural identification of photo-induced pyrimidine dimers in nucleic acids (for a review, see Ref. [6]). In the 1980's, emphasis has been placed on the preparation of oligonucleotides containing a pyrimidine dimer. The cyclobutyl structure of these pyrimidine photoproducts was ascertained on the basis of ^{1}H NMR analysis and X-ray diffraction. One of the major goals of most of these studies was the determination of the conformational changes induced in various DNA model compounds by the presence of a bulky pyrimidine dimer lesion. A likely hypothesis is that the distortion provoked

1.1 Pyrimidine Dimer

by a bulky lesion such as $Pyr <> Pyr$ within a DNA chain may serve as a signal for damage recognition by repair enzymes. In 1986, Pedrini et al. suggested on the basis of gel electrophoresis experiments that the presence of $Pyr <> Pyr$ within DNA chains would induce structural distortion as the result of concomitant unwinding and bending of the helix in the vicinity of the photolesion [7]. Oligonucleotides containing a pyrimidine dimer have also been the subject of a number of computational studies [8, 9, 10, 11] that have predicted that the lesion causes little bending of DNA, to bending as high as 27°. In 2002, Park et al. reported the crystal structure of a *cis-syn* thymine dimer in free (unbound) DNA decamer at a resolution of 2.5 Å, providing a clear picture of the effects of dimer formation on DNA structure and base pairing [12]. Most notably, the dimer is found to bend DNA by 30°, to induce an unexpected minor groove and in addition one of the two thymines displays weakened hydrogen-bonding with the complementary adenine. Our QM/MM theoretical investigations on self-repair of the thymine dimer in Chapter 4 are based on this structure.

1.1.1 Pyrimidine Dimer Repair

A curious discovery was made more than 50 years ago [13, 14]. Bacteria given a lethal dose of UV radiation can often be saved by irradiating with visible or near UV light. This photoreactivation, which permits many bacteria to survive, results from the action of the enzyme DNA photolyase, which absorbs light maximally at 380 nm and carries out a photochemical reversal of $Pyr <> Pyr$ in DNA, cutting the pyrimidine-pyrimidine covalent bonds [15]. In 1984, Sancar et al. identified photolyase as a flavoprotein containing two noncovalently bound chromophores [16]. One chromophore is the fully reduced flavin-adenine dinucleotide ($FADH^-$), the catalytic cofactor that carries out the repair reaction upon excitation by either direct photon absorption or energy transfer from the second chromophore, which is an antenna pigment that harvests sunlight and enhances repair efficiency. In 1987, Sancar et al. proposed a model for the catalytic reaction: the excited flavin cofactor transfers an electron to $Pyr <> Pyr$ causing its reversal to two pyrimidines [17]. The reduced chromophore is regenerated at the end of the photochemical step thus enabling the enzyme to act catalytically. At that time, Sancar had difficulty proving his scheme because he could not experimentally capture the proposed radical intermediates. Nearly 20 years after he first proposed the reaction mechanism, instrumentation has improved to a point where the mechanism can be demonstrated. In 2005, Sancar captured the elusive photolyase radicals he had chased for, thus providing direct observation of the photocycle for $Pyr <> Pyr$ repair [18]. Nevertheless, one subtle issue that remains unresolved is whether the splitting of the dimer is asynchronously concerted or sequential.

The binding of $Pyr <> Pyr$ to the enzyme active site to predict the structure of the enzyme-substrate complex has also been widely studied, and two different binding models have emerged from computational investigations. In the model developed independently by Rösch and co-workers [19] and Wiest and co-workers [20] in 1999, the dimer is $\approx$ 10 Å away from the redox active FADH cofactor and no direct contact is predicted. Thus, these studies suggest an electron transfer mediated by the π-systems of close tryptophans. The second model, developed by Stuchebrukhov and co-workers in 2000, suggests that the dimer and the adenine moiety of FADH are within hydrogen bonding distance of each other, predicting that the flavin passes its electron directly to the lesion [21]. At the end of 2004, the first X-ray structure of a photolyase bound to its damaged DNA substrate was reported, confirming that the enzyme flips the lesion into an active-site cavity right next to the catalytic flavin cofactor [22]. This breakthrough enabled us to include the protein environment in our QM/MM simulations of the repair reaction (see Chapter 5).

Recent experimental investigations have shown evidence for a self-repair mechanism in DNA as well [23, 24]. Adjacent bases to dipyrimidine sites play a crucial role in controlling the levels of $Pyr <> Pyr$ by acting as transient electron donors to promote repair of the lesion. The discovery of this autocatalytic process sheds light on how evolution may have been possible in a primordial *RNA world*, where the atmosphere of the early Earth was believed to have been subjected to high levels of UV radiation.

1.1.2 Pyrimidine Dimer Formation

The determination of the mechanism of formation of the pyrimidine dimer by using various nucleic acid model compounds has been the subject of numerous investigations (for reviews, see Ref. [6]). In 1967, Sztumpf-Kulikowska proposed that photodimerization of pyrimidine nucleobases in dilute aqueous solutions in-

1.1 Pyrimidine Dimer

volves photoexcitation of a molecule to a singlet state followed by intersystem crossing and subsequent reaction of the resulting triplet pyrimidine with a second molecule in the ground state [25]. Two years later, Whillans et al. could observe the triplet state by flash photolysis in solutions of uracil [26]. Further support for the significant involvement of the triplet state in the formation of the pyrimidine dimer was provided by using specific triplet quenchers such as dienes [27] or oxygen [28, 26]. However, a different mechanism has been proposed in 1970 to explain the higher efficiency of dimerization of thymine when exposed to far-UV light in concentrated aqueous solutions [29]. Under these conditions, pyrimidine aggregates are produced as the result of van der Waals stacking. It was inferred that photodimerization in the stacked complexes predominantly involves a singlet excited state or a singlet excimer intermediate. Experiments in the solid state have shown that the yields and the stereochemistry of $Pyr <> Pyr$ solutions depend on the spacing and the orientation of the pyrimidine nucleobases in their ground state [30]. The high efficiency of photo-induced dimerization of thymine in frozen aqueous solutions has been explained in terms of a suitable parallel arrangement of neighboring crystals of the hydrated nucleobases [31]. Despite the intense work in this field, the nature of the excited-state precursors to $Pyr <> Pyr$ is still unclear. Moreover, the connection between pyrimidine-dimer yield and local DNA conformation is poorly understood. Very recent time-resolved measurements shed some light on these issues by showing that the thymine dimers are fully formed around 1 ps after UV excitation in polymeric DNA [32]. They concluded that the initial excited singlet state can decay to a dimer photoproduct along a nearly barrierless pathway if the nucleobases are properly oriented at the instant of light absorption. A few geometrical requirements for reaction to occur were suggested, such as base stacking which reduces the distance between the two thymine bases.

This thesis is organized as follows: in Chapter 2 the theoretical foundations of the applied computational methods are introduced. Chapter 3 presents an *ab initio* study of gas-phase thymine dimer repair and formation. Chapter 4 presents a QM/MM investigation of self-repair of the thymine dimer. Chapter 5 deals with the repair of the thymine dimer by DNA photolyase. Chapter 6 provides a mechanistic picture of the thymine dimerization process in DNA. Chapter 7 draws the conclusions of this work and gives an outlook of possible future developments related to this work.

Chapter 2

Computational Methods

In less than 50 years, the field of computational chemistry has gone from being essentially nonexistent to being an active counterpart in experimental investigations, with high-performance computing, clever algorithmic implementations, and information technology dramatically influencing methods development and performance.

This chapter briefly summarizes the computational chemistry techniques used in this thesis.

Molecular Dynamics (MD) simulations provide atomic details of the structures and motions of a classical many-body system and hence allow for computing its dynamic and thermodynamic properties. In this context, the word *classical* means that the nuclear motion of the constituent particles obeys the laws of classical mechanics. Molecular dynamics is a multidisciplinary method. Its laws and theories stem from mathematics, physics, and chemistry, and it employs algorithms from computer science and information theory. It was originally conceived within theoretical physics in the late 1950's [33, 34], but is nowadays applied to various areas of science such as materials science and biochemistry.

The time evolution of a molecular system during MD simulations is described by Newton's equation of motion

$$\boldsymbol{F}_i = -\frac{\partial V}{\partial \boldsymbol{R}_i} = M_i \frac{d^2 \boldsymbol{R}_i}{dt^2} \qquad (2.1)$$

where $\boldsymbol{F}_i$ is the force acting on atom i with position $\boldsymbol{R}_i$ and mass M_i, and V is the potential energy of the system.

Computing the classical trajectory exactly would require to solve a system of 3N second order differential equations, where N is the number of atoms. In practice, these equations are never solved exactly but rather approximated by a suitable algorithm based on time discretization. The size of the time step must be chosen small enough to avoid discretization errors (i.e. much smaller than the fastest vibrational frequency in the system). Typical timesteps are in the order of 1 femtosecond for classical MD. A commonly used integration algorithm is the velocity Verlet scheme [35], which uses a Taylor expansion truncated beyond the quadratic term for the coordinates

$$R(t + \Delta t) = R(t) + v(t)\Delta t + \frac{F(t)}{2M}\Delta t^2. \qquad (2.2)$$

The update for the velocities is given by

$$v(t + \Delta t) = v(t) + \frac{F(t + \Delta t) + F(t)}{2M}\Delta t. \qquad (2.3)$$

If the system is isolated from changes in moles (N), volume (V) and energy (E), the ensemble generated by an MD simulation is the microcanonical ensemble (NVE). A microcanonical molecular dynamics trajectory may be seen as an exchange of potential and kinetic energy, with total energy being conserved. However, most chemical and biological processes occur at constant temperature and constant pressure. By coupling the system to a thermostat and/or by introducing pressure coupling, a canonical ensemble (NVT) or a isothermal-isobaric ensemble (NPT), respectively, can be sampled in MD simulations. Fore more details see Ref. [36].

2.1 Classical Molecular Dynamics

In classical MD the potential energy of a system of particles is described by an empirical force field (potential function). The parameters that enter the function are fitted to experimental or higher level computational data. Several force fields have been developed such as AMBER, GROMOS and CHARMM, which have been primarily parametrized for molecular dynamics of macromolecules, although they are also commonly applied for energy minimization. The basic functional form of a force field encapsulates both bonded terms relating to atoms that are linked by covalent bonds, and nonbonded (also called "noncovalent") terms describing the long-range electrostatic and van der Waals forces. The specific decomposition of the terms depends on the force field. Here, we use the AMBER8/parm 99 force field [37], which is of the form

$$V = \sum_{bonds} K_R (R - R_{eq})^2 + \sum_{angles} K_\theta (\theta - \theta_{eq})^2 + \sum_{dihedrals} \frac{V_n}{2}[1 + \cos(n\phi - \gamma)]$$
$$+ \sum_{i<j} \left[\frac{A_{ij}}{R_{ij}^{12}} - \frac{B_{ij}}{R_{ij}^6} + \frac{q_i q_j}{\epsilon R_{ij}}\right]. \qquad (2.4)$$

The bonded term includes three different contributions representing bond stretching, bond-angle bending and dihedral-angle torsion. The non-bonded term includes a first contribution describing the van der Waals interactions and a second one describing the electrostatic interactions between atoms i and j (Coulomb term). The van der Waals term

$$V_{vdW} = \sum_{i<j} \left[\frac{A_{ij}}{\boldsymbol{R}_{ij}^{12}} - \frac{B_{ij}}{\boldsymbol{R}_{ij}^{6}} \right] \qquad (2.5)$$

describes the repulsive force at short ranges (the result of overlapping electron orbitals, referred to as Pauli repulsion, decaying with $\boldsymbol{R}_{ij}^{-12}$) and the attractive force at long ranges (van der Waals force, or dispersion force, decaying with $\boldsymbol{R}_{ij}^{-6}$). Finally the electrostatic term

$$V_{el} = \sum_{i<j} \left[\frac{q_i q_j}{\epsilon \boldsymbol{R}_{ij}} \right] \qquad (2.6)$$

is due to internal distribution of the electrons, creating positive and negative parts of the molecule. The Coulomb interaction is a long-range interaction and the sum in Eq. (2.6) converges very slowly. Therefore different algorithms have been developed for a fast and accurate treatment of electrostatic interactions, based on Ewald summations. Particle mesh Ewald (PME) [38], smooth particle mesh Ewald (SPME) [39] and particle-particle/particle-mesh Ewald (P3M) [40] algorithms are widely used in classical MD programs.

Despite its overwhelming success, the bias that is necessarily introduced when the interatomic interactions are described through empirical potentials implies serious drawbacks. Apart from a lack of description of changes in chemical bonding, the transferability of the force field parameters can often be questioned. Moreover, induced polarization and charge transfer effects are difficult to implement and are currently neglected in most MD studies. As a rule of thumb, a first-principles description is necessary when the chemistry of the system plays an important role, e.g. when there is making and breaking of chemical bonds, changing environments, variable coordination, etc. If this is not the case, then it is better to use classical MD, which allows for much longer simulations of much larger samples, leading to a significant improvement in the statistics required to estimate thermodynamic quantities.

2.2 Ab initio Molecular Dynamics

Ab initio Molecular Dynamics (AIMD) schemes overcome the above mentioned limitations of classical force field simulations. The fact that the trajectories are realistic is a consequence of the first-principles description of the acting forces, which is achieved at the expense of introducing the electronic component explicitly, within the adiabatic approximation. Density functional theory (DFT) is only one of the possible realizations of a first-principles calculation, but it is the most widely used. The advantage of DFT is that its computational cost, at least within local or semi-local approximations like LDA and GGA, is significantly lower than Hartree-Fock based wavefunction methods.

AIMD calculations are typically performed by using a plane-wave expansion of the DFT (Kohn-Sham) orbitals [41]. Although plane waves have the advantage of simplicity, they lead to $n^2 M$ scaling behaviour in computational cost with system size, where M is the number of plane waves. Formulation of AIMD in terms of a mixed approach based on Gaussians and plane waves has been proposed [42] and implemented in the CP2K code [43, 44]. The use of both localized and plane wave basis sets will potentially lead to linear scaling AIMD methodology in near future.

We will first give a short introduction to DFT before describing different MD schemes.

2.2.1 Density Functional Theory

Traditional methods in electronic structure theory, in particular Hartree-Fock theory and its descendants, are based on the complicated many-electron wavefunction. The main objective of density functional theory is to replace the many-body electronic wavefunction with the electronic density as the basic quantity. Whereas the many-body wavefunction is dependent on 3N variables, three spatial variables for each of the N electrons, the density is only a function of three spatial coordinates and is a simpler quantity to deal with both conceptually and practically.

Modern DFT is based on two theorems introduced by Hohenberg and Kohn [45]. The first theorem states that the external potential $\nu_{ext}(\mathbf{r})$ is uniquely determined by the ground state density ρ_0 up to a constant

$$\rho_0 \mapsto \nu_{ext}(\boldsymbol{r}). \tag{2.7}$$

Since the number of electrons (N_e) is uniquely defined by the electron density, $N_e = \int \rho_0(\boldsymbol{r})d\boldsymbol{r}$, ρ_0 determines the full Hamiltonian and therefore implicitly all properties of the system. The first theorem allows us to write the total energy as a functional of the electron density in the following way:

$$E_0 = E_0[\rho_0] \tag{2.8}$$

$$E_0[\rho_0] = T[\rho_0] + \int \rho(\boldsymbol{r})v_{ext}(\boldsymbol{r})d\boldsymbol{r} + V_{ee}[\rho_0] \tag{2.9}$$

where $T[\rho_0]$ is the kinetic energy and $V_{ee}[\rho_0]$ is the electron-electron interaction energy. The exact from of the terms describing the kinetic energy and the electron interaction energy are not known. Thus, the energy cannot be determined.

The second theorem introduces the energy variational principle. It states that there exists a universal functional that yields the lowest energy if and only if the input density is the true ground state density, ρ_0

$$E[\tilde{\rho}] \geq E[\rho_0]. \tag{2.10}$$

In 1965, Kohn and Sham suggested an avenue for how the unknown energy functional can be approximated [46]. They proposed to express the kinetic energy as the kinetic energy of a fictitious reference system s of n non-interacting electrons

$$T_s[\rho_s] = \sum_i^n \langle \phi_i^{KS} | -\frac{1}{2}\nabla^2 | \phi_i^{KS} \rangle. \tag{2.11}$$

The connection of this artificial system to the one we are really interested in is established by choosing the effective potential $\nu_{ext,s}$ such that the density resulting from the summation of the moduli of the squared orbitals exactly equals the ground state density of our real target system of interacting electrons

$$\rho_s(\boldsymbol{r}) = \rho_0(\boldsymbol{r}) = \sum_i^n |\phi_i^{KS}(\boldsymbol{r})|^2 \tag{2.12}$$

2.2 Ab initio Molecular Dynamics

where $\phi_i^{KS}(\boldsymbol{r})$ are the orthonormal Kohn-Sham orbitals. The expression of the electron density and the kinetic energy is exact for a one determinant wave function of a system of non-interacting electrons. The difference in kinetic energy and in electronic interaction energy between the reference system and the real system is

$$\Delta T[\rho] \equiv T[\rho] - T_s[\rho] \tag{2.13}$$

$$\Delta V_{ee}[\rho] \equiv V_{ee}[\rho] - \frac{1}{2} \int \int \frac{\rho(\boldsymbol{r_1})\rho(\boldsymbol{r_2})}{|\boldsymbol{r_1} - \boldsymbol{r_2}|} d\boldsymbol{r_1} d\boldsymbol{r_2}. \tag{2.14}$$

Insertion in the Hohenberg-Kohn Eq. (2.9) yields

$$E^{KS}[\rho] = \int \rho(\boldsymbol{r})\nu_{ext}(\boldsymbol{r})d(\boldsymbol{r}) + T_s[\rho] + \frac{1}{2} \int \int \frac{\rho(\boldsymbol{r_1})\rho(\boldsymbol{r_2})}{|\boldsymbol{r_1} - \boldsymbol{r_2}|} d\boldsymbol{r_1} d\boldsymbol{r_2} + E_{xc}[\rho] \tag{2.15}$$

with

$$E_{xc}[\rho] \equiv \Delta T[\rho] + \Delta V_{ee}[\rho]. \tag{2.16}$$

The exchange-correlation functional $E_{xc}[\rho]$ represents the non-classical part of the electronic interaction energy and the difference in kinetic energy between the reference system and the real system. The Kohn-Sham orbitals are found by minimization of Eq. (2.15) under the constraint that $\langle \phi_i | \phi_j \rangle = \delta_{ij}$. This results into the Kohn-Sham equations

$$\left\{ -\frac{1}{2}\nabla^2 + \nu_{exc} + \int d\boldsymbol{r}' \frac{\rho(\boldsymbol{r}')}{|\boldsymbol{r} - \boldsymbol{r}'|} + \nu_{xc} \right\} \phi_i(\boldsymbol{r}) = \epsilon_i \phi_i(\boldsymbol{r}), \tag{2.17}$$

which have to be solved self-consistently.

If only the correct expression for the exchange-correlation potential,

$$\nu_{exc}(\boldsymbol{r}) \equiv \frac{\delta E_{xc}[\rho(\boldsymbol{r})]}{\delta \rho(\boldsymbol{r})}, \tag{2.18}$$

was known, solving Eq. (2.17) would be equivalent to solving the exact electronic Schrödinger equation. Unfortunately, the exact exchange-correlation potential is unknown and much effort has been and is being devoted to find good approximations to ν_{exc}. Therefore, the quality of the electronic structure calculation depends on the quality of the approximation used for E_{exc}.

The local density approximation LDA is the simplest approximation for this functional, it is local in the sense that the electron exchange and correlation energy at any point in space is a function of the electron density at that point only

$$E_{xc}^{LDA}[\rho] = \int \rho(\boldsymbol{r})\epsilon_{xc}(\rho)d\boldsymbol{r}, \qquad (2.19)$$

where $\epsilon_{xc}(\rho)$ is the sum of the exchange and correlation energy of electrons in a homogeneous electron gas of density ρ. LDA yields good results for solid state systems, but for molecules the homogeneous electron gas approximation is in general too crude. In the early eighties, the first successful extensions to the purely local approximation was developed. The logical first step in that direction was the suggestion of using not only the information about the density $\rho(\boldsymbol{r})$ at a particular point $\boldsymbol{r}$, but to supplement the density with information about the gradient of the charge density, $\nabla \rho(\boldsymbol{r})$ in order to account for the non-homogeneity of the true electron density. Many functionals have been developed in the framework of the generalized gradient approximation (GGA). In this work, we use the BLYP functional due to Becke [47] for the exchange part and due to Lee, Yang, and Parr [48] for the correlation part and the PBE functional due to Perdew, Burke, and Ernzerhof [49]. A significant improvement of the accuracy of DFT calculations was achieved by introducing the so-called hybrid functionals, which include to some extent "exact exchange" from Hartree-Fock theory in addition to standard GGA. In particular the B3LYP functional from Becke [50] soon developed into the most popular hybrid functional. However, these functionals are prohibitively time-consuming in the CPMD code because of the computational cost of calculating the two-electron integral of non-local exchange along with the plane wave basis set.

2.2.2 Born-Oppenheimer Molecular Dynamics

In Born-Oppenheimer Molecular Dynamics (BOMD), the static electronic structure problem is solved at every MD step given the set of fixed nuclear positions at that instant of time. Thus, the electronic structure part consists in solving the

2.2 Ab initio Molecular Dynamics

time-independent Schrödinger equation, while the nuclei are propagating via classical molecular dynamics. Thus, the time-dependence of the electronic structure is a consequence of nuclear motion. The BOMD method is defined by

$$M_I \ddot{R}_I(t) = -\nabla_I \min_{\Psi_0}\{\langle\Psi_0|H_e|\Psi_0\rangle\} \quad (2.20)$$

$$E_0\Psi_0 = H_e\Psi_0. \quad (2.21)$$

According to Eq. (2.20), the minimum of $\langle H_e \rangle$ has to be reached at each BOMD step. Since the accuracy of the forces depends linearly on the accuracy of the minimization of the Kohn-Sham energy, the wave function has to be tightly converged at each step.

2.2.3 Car-Parrinello Molecular Dynamics

The Car-Parrinello approach is closely related to BOMD since the ions are also propagated classically. The fundamental difference is that the orbitals are no longer optimized at every time step but treated and propagated like classical objects, correspondingly being assigned a fictitious mass (μ) and temperature [51]. It could be shown that the adiabatic separation of the BO-approximation is also conserved for this approach [41]. In order to maintain this adiabaticity condition, it is necessary that the fictitious mass of the electrons is chosen small enough to avoid a significant energy transfer from the ionic to the electronic degrees of freedom. This small fictitious mass in turn requires that the equations of motion are integrated using a smaller time step (0.1-0.2 fs) than the ones commonly used in BOMD (0.5-1 fs). Hence, the computational bottleneck of BOMD, i.e the wavefunction optimization at each time step, can be circumvented within Car-Parrinello Molecular Dynamics (CPMD). More details on CPMD can be found in [41].

The BO scheme will be mostly used in this thesis since a highly efficient wavefunction optimization procedure, namely the orbital transformation technique [52], has been implemented in the CP2K code. This method allows to use BOMD without any computational overhead with respect to CPMD. Indeed, test calculations in our group showed that the number of SCF-steps required for one

BO-step is similar to the number of CP-steps required for the same time.

2.3 Hybrid QM/MM Molecular Dynamics

Even with present-day hardware and the most efficient linear scaling method, many of the complex systems that are of current interest in biology and nanotechnology are too large for a straightforward application of fully ab initio methods. However, the properties to be adressed are often local in nature, such as the chemical reactivity of specific sites. In these cases, a quantum mechanical description is necessary only for a small number of atoms around the site of interest, the rest of the system affects the local properties only via long range electrostatic interactions and geometrical constraints. For this class of problems the so-called quantum mechanical/molecular mechanical (QM/MM) approach offers a satisfactory compromise between accuracy and computational efficiency. The basic strategy for this approach was laid out in a seminal paper by Levitt and Warshel [53]. The system is divided in two regions, one treated within the *ab initio* framework (QM region), and the second treated by a classical force field (MM region), where the total energy is given by

$$E_{TOT}(\bm{r}_\alpha, \bm{r}_a) = E^{QM}(\bm{r}_\alpha) + E^{MM}(\bm{r}_a) + E^{QM/MM}(\bm{r}_\alpha, \bm{r}_a) \qquad (2.22)$$

where E^{QM} is the pure quantum energy, E^{MM} is the classical energy, and $E^{QM/MM}$ represents the mutual interaction energy of the two subsystems. These energy terms depend parametrically on the coordinates of the quantum nuclei ($\bm{r}_\alpha$) and classical atoms ($\bm{r}_a$). The interaction energy term $E^{QM/MM}$ contains all non-bonded contributions between the QM and the MM subsystems and in a DFT framework is expressed as

$$E^{QM/MM}(r_\alpha, r_a) = \sum_{a \in MM} q_a \int \frac{\rho(r, r_\alpha)}{|r - r_a|} dr + \sum_{\substack{a \in MM \\ \alpha \in QM}} v_{VdW}(r_\alpha, r_a) \qquad (2.23)$$

where $\bm{r}_a$ is the position of the MM atom a with charge q_a, $\rho(\bm{r}, \bm{r}_\alpha)$ is the total (electronic plus nuclear) charge density of the quantum system, and $\nu_{VdW}(\bm{r}_\alpha, \bm{r}_a)$ is the van der Waals interaction between classical atom a and quantum atom α. In the employed QM/MM scheme, the van der Waals term is simply taken from the classical force field. Furthermore, the use of dispersion-corrected atom-centered potentials (DCACPs) provides an alternative approach to include London dispersion forces within the framework of Kohn-Sham density functional theory without incurring an unaffordable computational overhead [54]. The implementation of the electrostatic $E^{QM/MM}$ term is non-trivial due to (i) the electron spill-out effect because of the missing Pauli repulsion between the electrons and the positively charged nearby MM atoms and (ii) the high computational cost. The first problem is overcome by suitably modifying the Coulomb term at short range. Computational efficiency is achieved by using a multipolar expansion of the QM charge density to compute the long-range Coulomb interaction, which drastically reduces the number of operations to be performed. More details on this QM/MM scheme implemented in the CPMD code can be found in Ref. [55].

Recently, a novel real space multigrid approach that handles Coulomb interactions very efficiently has been implemented in the CP2K code [56]. This scheme cuts the cost of the coulombic interaction evaluation between the QM and the MM parts by 2 orders of magnitude with respect to the plane wave-based implementation.

2.4 Metadynamics

In order to observe "rare events" such as chemical reactions at moderate temperature, we will adopt the metadynamics (MTD) methodology [57, 58] throughout this thesis. This method has been successfully applied in combination with ab initio MD to the study of various types of reactive systems in which substantial changes to the electronic structure were expected [58, 59, 60, 61, 62]. It has been shown that this method is capable of providing accurate free energy information [63].

The subspace for which we wish to boost the sampling is defined by selecting a set of collective variables (CVs) that can distinguish the different states of the system under investigation. These variables are associated with some selected collective motions of the system, which describe the desired reaction path (e.g. stretching, bending, torsion, coordination numbers, or other more general coordinates). The meta-trajectory is determined by integrating the equations of motion derived from an extended Lagrangian [58] of the form

$$\mathcal{L} = \mathcal{L}_{BO} + \sum_\alpha \frac{1}{2} M_\alpha \dot{s}_\alpha^2 - \sum_\alpha \frac{1}{2} k_\alpha [S_\alpha(\mathbf{R}) - s_\alpha]^2 - V(t, \mathbf{s}) \qquad (2.24)$$

in which the additional dynamic variables s_α define the dynamics in the reduced space of the CV. The first term $\mathcal{L}_{BO}$ is the Born-Oppenheimer Lagrangian, which drives the electronic and ionic dynamics. The second is the fictitious energy of the new dynamic variables. The third term is a harmonic restraint potential that couples the meta-trajectory to the standard MD trajectory through the instantaneous values of the CVs, S_α. The fictitious mass M_α and the coupling constant k_α determine the frequency of the fluctuations of the meta-trajectory. The last term $V(t, \mathbf{s})$ ($\mathbf{s}$ is the vector of s_α) is the history-dependent potential acting on the fictitious particles, and its role is to enhance the sampling of the configurational space. It is constructed "on-the-fly" by the accumulation of Gaussian hills deposited at regular time intervals in order to reduce the probability of visiting again the same configurations, resulting in accelerated barrier crossing. At convergence, that is, when the available wells have been completely filled by the accumulated potential, the explored free energy surface (FES) can be reconstructed from $V(t, \mathbf{s})$ [58]. More details about the optimal choice of the time-dependent potential and of the other metadynamics parameters can be found in previous publications [58, 59, 60, 61, 62].

2.5 Excited-State Calculations

Two DFT-based approaches, namely the restricted open-shell Kohn-Sham (ROKS) formalism and time-dependent density DFT (TDDFT), were used in this thesis to descibe excited states.

The ROKS algorithm is a self-consistent scheme inspired by the Ziegler-Rauk-Baerends "sum methods" [65] that allows for MD simulations in the first excited singlet state S_1 [64]. Promoting one electron from the highest occupied molecular orbital (HOMO) to the lowest unoccupied molecular orbital (LUMO) in a closed-shell system leads to four different excited wavefunctions or determinants (Figure 2.1a). Two states $|t_1\rangle$ and $|t_2\rangle$ are energetically degenerate triplets t whereas

2.5 Excited-State Calculations

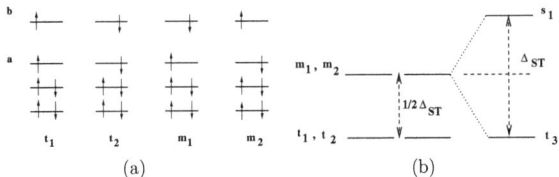

Figure 2.1: (a) Four different determinants resulting from promoting one electron from HOMO to LUMO and (b) two mixed states $|m_1\rangle$ and $|m_2\rangle$ can be combined to yield a triplet state $|t_3\rangle$ and the singlet state $|s_1\rangle$. The singlet-triplet splitting Δ_{ST} correponds to twice the splitting between triplet and mixed states. (Figure adapted from [64].)

the two states $|m_1\rangle$ and $|m_2\rangle$ are not eigenfunctions of the total spin operator. However, they can be combined to yield another triplet state $|t_3\rangle$ and the desired first excited singlet state $|s_1\rangle$ (Figure 2.1b). The total energy of the S_1 state is then given by

$$E_{S1} = 2E_m^{KS} - E_t^{KS}, \qquad (2.25)$$

and the associated S_1 wavefunction is given by

$$|s_1[\{\phi_i\}]\rangle = \sqrt{2}|m[\{\phi_i\}]\rangle - |t[\{\phi_i\}]\rangle, \qquad (2.26)$$

where $\{\phi_i\}$ denotes the complete set of orbitals. Using the definitions of the exchange-correlation potentials for α and β spin, respectively,

$$\nu_{xc}^\alpha = \frac{\delta E_{xc}[\rho^\alpha, \rho^\beta]}{\delta \rho^\alpha}, \qquad (2.27)$$

$$\nu_{xc}^\beta = \frac{\delta E_{xc}[\rho^\alpha, \rho^\beta]}{\delta \rho^\beta}, \qquad (2.28)$$

the corresponding Kohn-Sham equations are obtained by varying Eq. (2.25). The equation for the doubly occupied orbitals reads

$$\left\{ -\frac{1}{2}\nabla + V_H + \nu_{ext}(\boldsymbol{r}) \right.$$
$$+ \nu_{xc}^\alpha[\rho_m^\alpha(\boldsymbol{r}), \rho_m^\beta(\boldsymbol{r})] + \nu_{xc}^\beta[\rho_m^\alpha(\boldsymbol{r}), \rho_m^\beta(\boldsymbol{r})]$$
$$\left. - \frac{1}{2}\nu_{xc}^\alpha[\rho_t^\alpha(\boldsymbol{r}), \rho_t^\beta(\boldsymbol{r})] - \frac{1}{2}\nu_{xc}^\beta[\rho_t^\alpha(\boldsymbol{r}), \rho_t^\beta(\boldsymbol{r})] \right\} \phi_i(\boldsymbol{r}) = \sum_{j=1}^{n+1} \Lambda_{ij}\phi_j(\boldsymbol{r}) \qquad (2.29)$$

and the two different equations for the two singly-occupied open-shell orbitals a and b, respectively, read

$$\left\{ \frac{1}{2}\left[-\frac{1}{2}\nabla + V_H + \nu_{ext}(\boldsymbol{r}) \right] \right.$$
$$\left. + \nu_{xc}^\alpha[\rho_m^\alpha(\boldsymbol{r}), \rho_m^\beta(\boldsymbol{r})] - \frac{1}{2}\nu_{xc}^\alpha[\rho_t^\alpha(\boldsymbol{r}), \rho_t^\beta(\boldsymbol{r})] \right\} \phi_a(\boldsymbol{r}) = \sum_{j=1}^{n+1} \Lambda_{aj}\phi_j(\boldsymbol{r}), \qquad (2.30)$$

and

$$\left\{ \frac{1}{2}\left[-\frac{1}{2}\nabla + V_H + \nu_{ext}(\boldsymbol{r}) \right] \right.$$
$$\left. + \nu_{xc}^\beta[\rho_m^\alpha(\boldsymbol{r}), \rho_m^\beta(\boldsymbol{r})] - \frac{1}{2}\nu_{xc}^\alpha[\rho_t^\alpha(\boldsymbol{r}), \rho_t^\beta(\boldsymbol{r})] \right\} \phi_b(\boldsymbol{r}) = \sum_{j=1}^{n+1} \Lambda_{bj}\phi_j(\boldsymbol{r}). \qquad (2.31)$$

These equations can be solved by iterative diagonalization or by minimization. Direct minimization of the total energy functional using an algorithm for orbital-dependent functionals [66] has been implemented in the CPMD code.

TDDFT was only employed to compute vertical excitation energies since its use within an AIMD scheme requires an important computational effort. TDDFT is an extension of DFT to the time-dependent domain and as such the key quantity is the electronic density, at least in the first formulation of the method [67]. A detailed review can be found in Ref. [68].

In a nutshell, this strategy employs the fact that the frequency dependent linear response of a finite system with respect to a time-dependent perturbation has discrete poles at the exact, correlated excitation energies of the unperturbed

2.5 Excited-State Calculations

system. To be more specific, the frequency dependent mean polarizability $\alpha(\omega)$ describes the response of the dipole moment to a time-dependent electric field with frequence $\omega(t)$. It can be shown that the $\alpha(\omega)$ are related to the electronic excitation spectrum according to

$$\alpha(\omega) = \sum_I \frac{f_I}{\omega_I^2 - \omega^2}. \tag{2.32}$$

Here ω_I is the excitation energy $E_I - E_0$ and the sum runs over all excited states I of the system. From Eq. (2.32) we see that the dynamic mean polarizability $\alpha(\omega)$ diverges for $\omega_I = \omega$, i.e., has poles at the electronic excitation energies ω_I. The residues f_I are the corresponding oscillator strengths. Translated into the Kohn-Sham scheme, the exact linear response can be expressed as the linear density response of a non-interacting system to an effective perturbation. The orbital eigenvalue differences of the ground state KS orbitals enter this formalism as a first approximation to the excitation energies, which are then systematically shifted towards the true excitation energies. In the TDDFT approach, only properties of the ground state - namely the ordinary Kohn-Sham orbitals and their corresponding orbital energies obtained in a regular ground state calculation - are involved. Hence, excitation energies are expressed in terms of ground state properties and the problem of whether DFT can be applied to excited states is most elegantly circumvented. All the systematic investigations published so far essentially agree that TDDFT provides accurate excitation energies that rival more sophisticated and much more costly wave function-based approaches, as long as we are dealing with low-energy transitions involving valence states.

Chapter 3

A Quantum Chemical Study of Thymine Dimer Repair and Formation

Abstract

As a first step toward modeling the photoinduced repair of a lesion in DNA, electronic structure calculations on the cleavage reaction of the thymine dimer have been carried out in the gas-phase at the GGA/DFT level. Upon electron uptake, a spontaneous cleavage of the C5-C5' bond occurs. According to metadynamics simulations, the activation energy of subsequent C6-C6' bond breaking amounts to 6 kcal/mol. Calculations of vertical excitation energies were performed as well to get a first insight into the formation of the thymine dimer.

3.1 Introduction

Cyclobutane pyrimidine dimers (CPD) are formed by the [2 + 2] cycloaddition reaction upon absorption of ultraviolet light (UV) [3]. This is the classic example of a photochemical reaction allowed by the rules of conservation of orbital symmetry (Woodward-Hoffmann rules) [69]. The same rules apply to the splitting of the dimer by [2 + 2] cycloreversion with far UV. This orbital symmetry-allowed photoreaction will, however, not take place since thymine dimers do not efficiently absorb light as they lack the conjugated π system of the original thymines. Instead, an electron transfer either from the enzyme or from adjacent bases to the dimer initiates splitting according to the proposed mechanism which is summarized in Figure 3.1 [18].

Splitting of the cyclobutane ring is not a photochemical but is a thermal reaction since CPD^- is not a photochemically excited species. Hence, it must follow the rules of the conservation of orbital symmetry for thermal reactions [69]. In fact, it can be shown that the conversion of a cyclobutane radical anion into ethene + ethene radical anions is also forbidden by the rules of conservation of orbital symmetry for thermal reactions [69]. Thus, it appears that the electron uptake lowers the splitting activation energy barrier allowing a "symmetry-forbidden" reaction to proceed very efficiently at ambient temperatures [70].

Several static gas-phase studies of the thymine dimer splitting mechanism have been performed [71, 72, 73, 74]. Two recent studies [72, 74] predicted a barrierless cleavage of the C5-C5' σ bond. However, there is no consensus on the value of the energy barrier for the breakage of the C6-C6' σ bond in the dimer radical anion. A barrier of 14.1 kcal/mol was found at the UHF/6-31G* level [73], whereas a much lower value (2.3 kcal/mol) was reported using B3LYP/6-311++G**//B3LYP/6-31+G* calculations [74].

Furthermore, we would like to point out that a gas-phase model is not expected to give an adequate representation of the repair reaction which occurs in solution or in the active site of DNA photolyase. Indeed, the valence-bound state of pyrimidines, which has the excess electron, can be stabilized by hydrogen-bonds [75, 76] that can be provided by the complementary adenines or by the residues in the active site of the enzyme. Saettel et al. have investigated the mechanism of the splitting of the uracil dimer radical anion hydrogen-bonded to three water molecules at the B3LYP/6-311++G**//B3LYP/6-31G* level of theory and found an asynchronously concerted mechanism, in which the C5-C5' bond first cleaves with an activation energy of 1.1 kcal/mol and the C6-C6' bond consecutively breaks barrierless [73]. Using a minimal model system with hydrogen-bonds, they demonstrated that the mechanism is different from the one disclosed by a gas-phase model. As will be discussed in Chapters 4 and 5, the inclusion of the DNA and the enzyme environments suggests an asynchronously concerted repair mechanism, in which a spontaneous C5-C5' bond cleavage upon electron uptake is followed by the C6-C6' bond cleavage with an upper bound to the activation energy of 2.5 kcal/mol. Overall, these results demonstrate the necessity of constructing a proper model system for the calculations of the relevant mechanism.

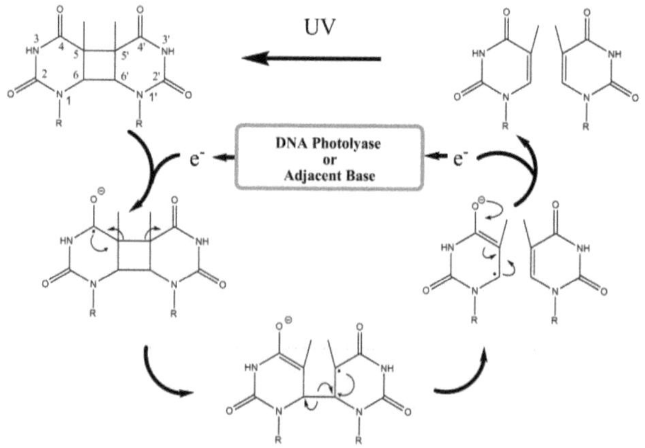

Figure 3.1: Reaction mechanism of splitting of thymine dimer. After electron transfer, the thymine dimer undergoes a [2 + 2] cycloreversion leading to reversion to base monomers.

Here, the repair reaction of the thymine dimer has been investigated in vacuo within the Car-Parrinello framework by metadynamics and by constrained geometry optimization. Moreover, vertical excitation energies have been computed for the constrained structures along the repair pathway to get a first insight into the formation of the thymine dimer. This photoreaction will be analyzed in the DNA environment in Chapter 6.

3.2 Methods

3.2.1 Computational Details

The thymine dimer was taken from a 2.5 Å resolution X-ray structure of a DNA decamer containing a cis-syn thymine dimer [12]. All calculations were performed at the DFT level of theory using the PBE functional [49] and the local spin-density approximation (LSD) as implemented in the CPMD code [77]. We use soft norm-conserving non-local Troullier-Martins pseudopotentials [78] and a 70 Ry energy cutoff for the plane-wave expansion of the wave function. The inherent periodicity in the plane-wave calculations is circumvented solving Poisson's equation for non-periodic boundary conditions [79]. A cubic cell with an edge length of 14 Å is sufficient to converge the energies and the geometries with respect to the cell size. MD simulations are carried out within the Car-Parrinello MD algorithm [51, 41] with a time step of 3 a.u. (0.07 fs) and a fictitious electron mass of 300 a.u..

The reaction profiles for the splitting of the thymine dimer radical anion and the neutral thymine dimer were investigated by constrained geometry optimization. Simulated annealing [80] was used to relax the structures starting with a temperature of 50 K until the convergence criterion for the energy was met (difference in energy per atom: $\Delta E = 2 \cdot 10^{-8}$ hartree/atom). The constraint coordinate is chosen as the distance between two dummy atoms placed at the bond midpoint between C5-C6 (D1) and C5'-C6' (D2).

The metadynamics technique [57, 58] was used to explore the free energy surface for the repair reaction of the thymine dimer radical anion. More details and references on this technique can be found in Section 2.4. The collective variables (CVs) were chosen as the C5-C5' and C6-C6' distances. For the two CVs the mass M_α and the coupling constant are 20 and 0.4 a.u., respectively. Gaussian-type hills 0.627 kcal mol^{-1} high and approximately 0.2 Å wide were used to build up the V$(t,\mathbf{s})$. The hills were added every 50 MD steps, and velocities of the fictitious particles (CVs) were scaled to maintain a temperature of 300 K. A repulsive potential wall was placed at 4.5 Å for the first (C5-C5' distance) and second (C6-C6' distance) CVs in order to limit the distance between the two thymines.

Calculations of the vertical excitation energies were carried out on the basis of the constrained structures along the neutral thymine dimer dissociation process using the TDDFT method [67] in the Tamm-Dancoff approximation [81].

3.3 Results and Discussion

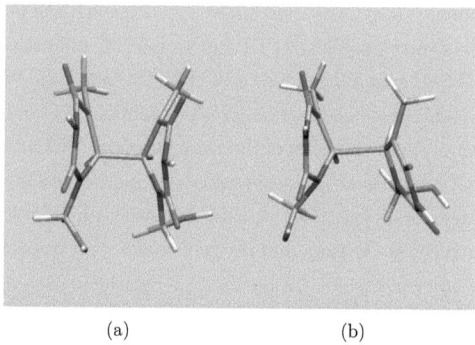

(a)   (b)

Figure 3.2: (a) Optimized structure of the thymine dimer radical anion and (b) transition-state structure found along the metadynamics trajectory using the shooting method.

Thymine Dimer Repair.

The potential energy surfaces for the splitting reactions of the thymine dimer radical anion and neutral thymine dimer are constructed by applying a constraint on a suitably chosen reaction coordinate, the distance between two dummy atoms placed at the bond midpoint between C5-C6 (D1) and C5'-C6' (D2). This distance constraint d(D1-D2) was varied in increments of 0.02 to 0.1 Å.

The addition of an electron to the thymine dimer leads to the spontaneous cleavage of the C5-C5' bond. The distance between C5 and C5' increases from 1.60 to 2.60 Å and the puckering angle rotates from 16 to 21° (Figure 3.2a). The C4-C5 and C4'-C5' bond lengths of the dimer radical anion are roughly 0.11 Å shorter than those in the neutral dimer, i.e. the C4-C5 bonds start to display double-bond character upon electron addition. Barrierless breaking of the C5-C5' bond was already predicted at the MP2 [72] and B3LYP [74] levels of theory. Subsequent cleavage of the C6-C6' bond requires ≈ 11 kcal/mol of activation

3.3 Results and Discussion

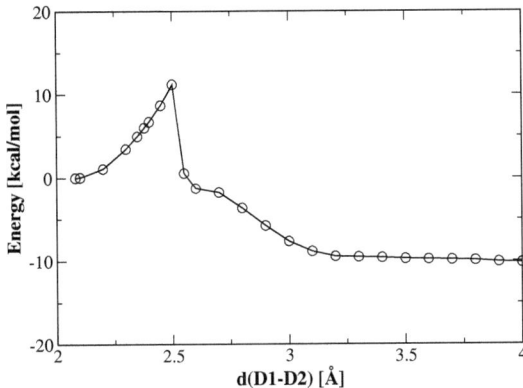

Figure 3.3: Potential energy curve for the thymine dimer radical anion splitting process in the ground-state as a function of the value of the constraint d(D1-D2).

energy (Figure 3.3). The two thymines are now ≈ 10 kcal/mol lower in energy than the ring-opened intermediate.

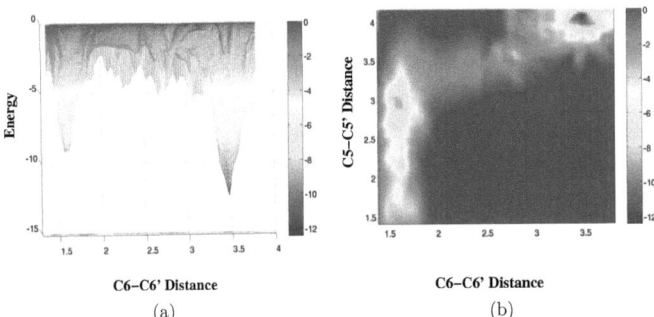

Figure 3.4: FES showing the repair process of the dimer radical anion as a function of the C6-C6' distance and (b) as a function of both C5-C5' and C6-C6' distances. The free energy is in kcal/mol and distances in Å.

The repair reaction of the thymine dimer radical anion was analyzed by metadynamics as well (Figure 3.4) and a free energy barrier of $\approx$ 6 kcal/mol was calculated. Moreover, the metadynamics simulation predicts that the two separated thymines lie $\approx$ 3 kcal/mol below the ring-opened intermediate. A previous DFT(B3LYP) study found that the splitting of the C6-C6' bond is slightly exothermic (2.4 kcal/mol), in agreement with our results [74].

In order to determine more precisely the transition state region along the metadynamics trajectory, we applied the shooting method [82] on several selected structures. The first basin of attraction corresponds to the ring-opened intermediate and the second one to the two separated thymines. The first basin is identified by the following values of CVs : d(C5-C5') = 2.6 $\pm$ 0.5 Å and d(C6-C6') = 1.6 $\pm$ 0.1 Å and the second one by d(C5-C5') = 4.2 $\pm$ 0 .2 Å and d(C6-C6') = 3.7 $\pm$ 0.3 Å. We could find one structure reaching the two basins of attraction with equal probability. This structure is characterized by a large puckering angle (67°) and its C5-C5' and C6-C6' bond lengths amount to 3.28 Å and 1.99 Å, respectively (Figure 3.2b).

Our values for the activation energy ($\approx$ 11 kcal/mol from constrained geometry optimization and $\approx$ 6 kcal/mol from metadynamics) lie in between the two values calculated in previous static calculations, i.e 14.1 kcal/mol at the UHF/6-31G* level [73] and 2.3 kcal/mol at the B3LYP level [74]. The difference in activation and reaction energies between the constrained geometry optimization and the metadynamics may be due to the entropic contribution. Rösch et al. estimated the entropic contribution to the reaction energy of the splitting of the uracil dimer radical anion to amount to about -14 kcal/mol in the gas-phase [71]. Our simulations point to a lower effect of the entropic contribution of $\approx$ -5 kcal/mol and of $\approx$ -7 kcal/mol for the reaction barrier and the reaction energy, respectively.

Finally, the energy profile for the neutral dimer indicates that the activation energy amounts to $\approx$ 40 kcal/mol and predicts that the reaction is exothermic by $\approx$ 25 kcal/mol (Figure 3.5). A similar exothermicity ($\approx$ -22 kcal/mol) was experimentally determined [70]. A high activation energy was expected since a neutral thymine dimer cannot revert to two thymines in a concerted thermal

3.3 Results and Discussion

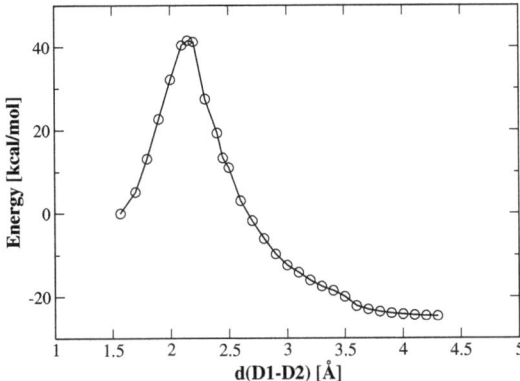

Figure 3.5: Potential energy curve for the neutral thymine dimer dissociation process in the ground-state as a function of the value of the constraint d(D1-D2).

process according to the Woodward-Hoffmann rules. However, Eriksson et al. found an activation energy even larger ($\approx$ 60 kcal/mol) at the B3LYP level [83], though they found a similar reaction energy ($\approx$ -20 kcal/mol). Different explanations are possible: (i) it was shown for the cycloaddition of ethylene to butadiene that PBE underestimates the activation energy up to 10 kcal/mol with respect to B3LYP [84], (ii) Eriksson et al. obtained the structures of the separated thymines, transition state and thymine dimer through unconstrained geometry optimization and characterized them by frequency calculations [83]. This mostly results in significant geometrical differences at the transition state as shown in Table 3.1. Our transition-state structure still has a C6-C6' bond, pointing to a stepwise mechanism, whereas their transition-state structure rather points to a concerted mechanism. Therefore, the reaction pathway that we have followed by increasing the D1-D2 distance involves a stepwise mechanism. This pathway is not symmetry forbidden as opposed to the concerted pathway and this may explain its lower activation energy.

If the splitting of the dimer radical anion is slightly exothermic, one might

	Separated thymines	Transition state	Thymine dimer	Ref
d(C5-C5')	4.23	2.66	1.60	this work
d(C5-C5')	4.18	2.34	1.59	[83]
d(C6-C6')	4.71	1.66	1.59	this work
d(C6-C6')	4.46	2.12	1.57	[83]
∠(C7-C5-C5'-C7')	132.5	11.3	24.6	this work
∠(C7-C5-C5'-C7')	35.3	6.8	27.6	[83]

Table 3.1: Comparison of key geometric parameters for selected structures along the ground-state pathway for the splitting of the neutral thymine dimer. Distances in Å and dihedral angles in degrees.

ask why, keeping in mind that our calculations suggest that the corresponding fragmentation of a neutral dimer into two neutral thymines would be exothermic by roughly 25 kcal/mol, evolution has favored an anionic reaction mechanism as opposed to a neutral? As already stated in the introduction, a thermally induced fragmentation of the neutral dimer requires a substantial amount of energy ($\approx$ 40 kcal/mol according to our calculations). Hence, the low-energy barrier to rupture the dimer radical anion ($\approx$ 6 kcal/mol according to the metadynamics simulation) could explain why nature has chosen an anionic reaction mechanism.

Thymine Dimer Formation.

We have first calculated the lowest vertical excitation energies of the thymine monomer in order to check our TDDFT(PBE) values against the ones reported in the literature. These values are summarized in Table 3.2. All previous calculations, either in the TDDFT framework or with quantum chemistry methods, predict the lowest transition to have an $n\pi^\star$ character in vacuo. We agree with this assignment since we find that the $S_0 \rightarrow S_1$ excitation is a symmetry-forbidden transition from the HOMO-1 to the LUMO (see Figure 3.6) with an extremely small oscillator strength. The next transition corresponds to a bright state and involves the HOMO and LUMO ($\pi \rightarrow \pi^\star$). The largest differences in excitation energies are found between TDDFT(PBE) and TDDFT(B3LYP), especially for the n $\rightarrow \pi^\star$ transition. The use of a hybrid functional (B3LYP) gives excitation energies much closer to experiment. In Chapter 6, excited-state metadynamics simulations will be performed to gain insight into the formation of thymine dimer in DNA. However, the use of hybrid functionals in plane wave-based ab

3.3 Results and Discussion

initio MD simulations is prohibitively expensive and thus we will employ pure density functionals. In conclusion, the assignment of the excited states agrees with previous calculations, but the TDDFT(PBE) energies are redshifted with respect to experiment, TDDFT(B3LYP) and high level ab initio calculations.

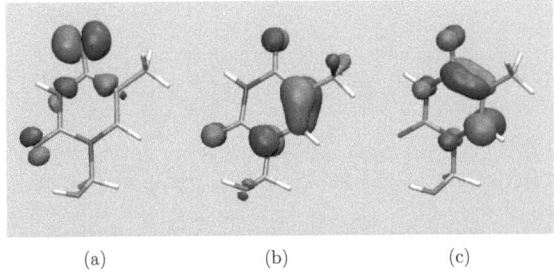

(a)   (b)   (c)

Figure 3.6: HOMO-1 (a), HOMO (b) and LUMO (c) orbitals of the thymine base in the optimized gas-phase geometry.

Next, we turn our attention to the thymine dimer. To the best of our knowledge, only one theoretical study has been carried out on the thymine dimer formation in vacuo [83, 85]. Eriksson et al. have computed the potential energy curves for a reaction proceeding via the lowest singlet and triplet excited states at the B3LYP level. It has been found that the thymine dimer formation can occur on both singlet and triplet excited potential energy surfaces in the gas-phase [83, 85].

Calculations of the vertical singlet and triplet excitation energies were performed on the constrained geometries along the neutral thymine dimer dissociation pathway (Figure 3.5). The lowest singlet and triplet transitions involve both the HOMO and LUMO. Figure 3.7 shows the HOMO and the LUMO for the structure at d(D1-D2) = 3.4 Å. The HOMO is distributed on the two stacked thymines, whereas the LUMO is localized on only one thymine. As we reduce d(D1-D2), the HOMO and the LUMO become equally distributed on the two thymines. For larger values of d(D1-D2), the molecular orbitals localize on different thymines.

Figure 3.8 shows the potential energy curve for the neutral thymine dimer

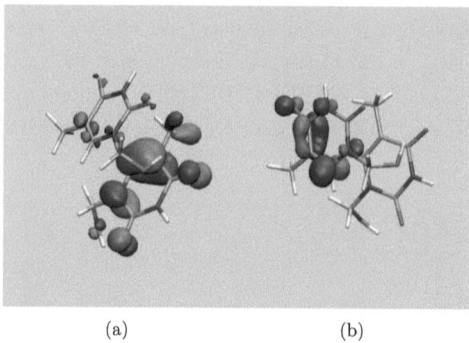

Figure 3.7: HOMO (a) and LUMO (b) orbitals of the stacked thymine bases at d(D1-D2) = 3.4 Å.

dissociation process (black) (see Figure 3.5) along with the corresponding singlet (red) and triplet (green) potential energy curves. We can see that the energy corresponding to the $S_0 \rightarrow S_1$ transition at the two dissociated thymines amounts to $\approx$ 3.20 eV (390 nm). This energy lies in near-UV in contrast to the energy obtained by Eriksson (4.61 eV) wich lies within the far-UV region known to induce the foramtion of the thymine dimer [83].

Even though the fact that we do not optimize the excited state geometries obviously means that the curves representing the photochemical reaction pathway should not, strictly speaking, be viewed as potential energy curves, we observe the presence of a "potential energy" well at a nuclear configuration corresponding to the transition state along the thermal reaction pathway. Starting from the two dissociated thymines, the system has to overcome an energy barrier of 46 kcal/mol and 22 kcal/mol on the singlet and triplet pathways, respectively, before the system reaches this local minimum. These barriers were much smaller (4 kcal/mol and 6 kcal/mol, respectively) in the study of Eriksson et al. [83]. In the case of the triplet excited state, the bottom of the well lies below the corresponding ground-state structure. This feature was not found by Eriksson et al. since the smallest gap between the triplet and the ground-state surfaces amounted to $\approx$ 14 kcal/mol along the energy profiles [85]. In the case of the singlet excited state, the gap between the two curves at the leakage channel amounts to $\approx$ 30 kcal/mol, which is twice as much as the value reported by Eriksson et al. ($\approx$ 16

3.3 Results and Discussion

kcal/mol) [83]. However, an early ab initio study of the photochemical disrotatory closure of butadiene to cyclobutene on the singlet excited state predicted a similar gap ($\approx$ 25 kcal/mol) [86] .

In conclusion, in contrast to the study of Eriksson et al. pointing to energetically favorable paths in terms of activation energy for the formation reaction on the singlet and triplet PES, we find rather high energy barriers to overcome before reaching the leakage channel. According to our results, the photochemical reaction pathway is thus not energetically accessible. These discrepancies may be explained (i) by the different functional used and (ii) by the different ground-state energy profiles. As it has already been mentioned in the last section about the splitting of the neutral thymine dimer, Eriksson et al. described a concerted reaction mechanism with an energy barrier of $\approx$ 60 kcal/mol, while we reported a stepwise mechanism with a barrier of only $\approx$ 40 kcal/mol. On both the singlet and triplet excited states, a potential energy "well" at the transition state of the ground-state PES was found. At this leakage channel, a gap of $\approx$ 30 kcal/mol between S_1 and S_0 was calculated.

Type		Method	Ref
$n\pi^\star$	4.04 (0.00002)	TDDFT(PBE)	this work
$\pi\pi^\star$	4.63 (0.10998)		
$n\pi^\star$	4.39 (0.00019)	CASPT2	[87]
$\pi\pi^\star$	4.88 (0.17)		
$n\pi^\star$	4.22 (0.000)	scaled CIS[b]	[88]
$\pi\pi^\star$	4.75 (0.505)		
$n\pi^\star$	4.76 (0.0001)	TDDFT(B3LYP)	[89]
$\pi\pi^\star$	5.17 (0.124)		
	4.5-4.7	exp.	[87]
	5.0-5.1	exp.	

Table 3.2: Lowest vertical excitation energies (eV) calculated for thymine, compared with selected calculations from litterature.[a]

[a]Results are only presented for the highest level of theory used in each study. Oscillator strengths are in parentheses. The canonical tautomer was studied for each base. [b] The scale factor is 0.697.

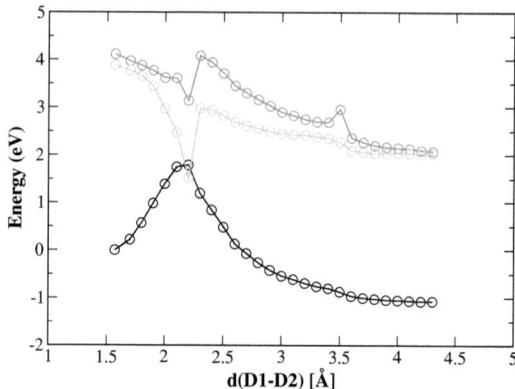

Figure 3.8: Potential energy curve for the neutral thymine dimer dissociation process in the ground-state (black) along with the corresponding singlet (red) and triplet (green) potential energy curves as a function of the value of the constraint d(D1-D2). The photochemical pathway solely involves the lowest excited triplet and singlet states of the different molecular arrangements along the reaction coordinate. Energies are given relative to the thymine dimer in the optimized gas-phase geometry.

3.4 Conclusion

The thymine dimer repair and formation reactions have been investigated in vacuo at the GGA/DFT level. In the repair process, a spontaneous C5-C5' bond cleavage is found upon electron uptake, in agreement with previous calculations [72, 74]. The activation and reaction energies of subsequent C6-C6' bond breaking amount to $\approx$ 6 kcal/mol and $\approx$ -3 kcal/mol, respectively, according to metadynamics simulations. These values are in good agreement with a previous study performed at the B3LYP level [74]. The energy differences with the constrained geometry optimization can be explained by the entropic contribution (5-7 kcal/mol). The splitting of the neutral thymine dimer requires an activation energy of $\approx$ 40 kcal/mol. However, this value does not match with the outcome of a B3LYP study by Eriksson et al. ($\approx$ 60 kcal/mol) [83]. This difference in energy may be related to the fact that we explored a different pathway via a stepwise mechanism as opposed to a concerted mechanism. In Chapter 6, our metadynamics simulations in DNA will reveal a concerted pathway over a similar barrier as that calculated by Eriksson et al..

Test calculations of the vertical excitation energies on the thymine monomer show that the use of a hybrid functional seems to highly improve the excitation energies. However, pure density functionals do not alter the state ordering. Our results do not predict energetically accessible paths on both singlet and triplet excited-states for the thymine dimer formation. Further investigations along a concerted pathway with a hybrid functional would be desirable in order to get a more accurate picture of the excited pathways for the formation reaction.

The next steps in our attempts to better understand the repair and formation processes will be the inclusion of the full environment, i.e. the DNA, and in the case of the repair reaction, the DNA photolyase as well. We will tackle these issues in the following chapters.

Acknowledgment

Computer resources were provided by the Swiss National Supercomputing Center (CSCS) and by the University of Zurich on the Matterhorn Cluster.

Chapter 4

Computational Evidence for Self-Repair of Thymine Dimer in Duplex DNA

4.1 Abstract

Formation of the thymine dimer is one of the most important types of photochemical damage in DNA responsible for several biological pathologies. Though specifically designed proteins (photolyases) can efficiently repair this type of damage in living cells, an autocatalytic activity of the DNA itself was recently discovered, allowing for a self-repair mechanism. In this article, we provide the first molecular dynamics study of the photoactivated repair mechanism of the thymine dimer using a QM/MM approach based on density functional theory (DFT) to describe the quantum region. A set of 7 statistically representative molecular dynamics trajectories is analyzed. Our calculations confirm the experimental results of a self-repair mechanism, predicting an asynchronously concerted process in which C5-C5' bond breaking is barrierless while C6-C6' bond breaking is characterized by a small free energy barrier. An upper bound of 2.5 kcal/mol for this barrier is estimated. The theoretical investigations confirm both the thermodynamical and kinetic feasibility of the self-repair process.

4.2 Introduction

Cyclobutane pyrimidine dimer is the most abundant damage caused to DNA by ultraviolet (UV) light [90], formed between two adjacent pyrimidine nucleobases, mainly thymine, via a [2 + 2] photocycloaddition. This lesion can lead to miscoding during DNA replication, which may cause mutations resulting in the development of skin cancer [91]. Recent studies directly connected the photochemical yield of thymine dimer (TT) formation to the deoxynucleotide (dN) sequence [92], DNA conformation [93, 32] and the binding of sequence-specific proteins to DNA [94]. The most ingenious strategy of a cell to repair this lesion is the light-driven photoreactivation catalyzed by DNA photolyase [95, 3, 96]. The commonly accepted model proposes that blue or near-UV light energy (300-500 nm) is absorbed by an antenna pigment which transfers the excitation energy to the reduced flavin coenzyme ($FADH^-$) of the photolyase. The excited $FADH^-$ then transfers an electron to the thymine dimer leading to destabilization of the C5-C5' and C6-C6' bonds, and thus boosting reversion to base monomers (see Fig. 4.1). However, this very elegant way of directly repairing damaged DNA,

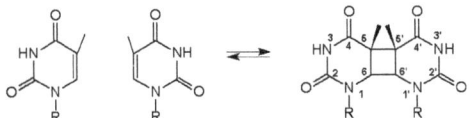

Figure 4.1: Formation and cycloreversion of the cis-syn thymine dimer.

used by many organisms, does not explain how evolution may have been possible in a primordial *RNA world*, where the self-replicating nucleic acid population was highly vulnerable to mutations via UV light-mediated pyrimidine dimer formation. A possible and surprising solution was recently proposed along with the discovery of autocatalytic properties of nucleic acids in photoreactivation reactions: a deoxyribozyme containing a guanine quadruplex can efficiently repair the thymine dimer via a mechanism reminiscent of DNA photolyase [23]. The guanine quadruplex was implicated as serving as a light-harvesting antenna, with photoreactivation of the thymine dimer proceeding possibly via electron donation from an excited guanine base. Nucleic acids themselves could thus have played a role in preserving the integrity of such an RNA world, autocatalyzing the splitting of the TT lesion sites. More recently, an experimental study [24] shed light on the self-repair mechanism of TT, emphasizing the importance of the nucleobases adjacent to dipyrimidine sites in controlling the levels of TT. Lower levels of TT were found in sites surrounded by guanine versus adenine, consistent with the fact that guanine is the nucleobase with the lowest oxidation potential, capable of acting as transient electron donor to promote repair of the TT lesion.

Many theoretical calculations regarding both the photochemical ring formation [97, 85] and the thymine repair mechanism [74, 71, 72, 73, 98] have been performed in order to elucidate the underlying molecular mechanism. Nonetheless, it is still unclear whether the mechanism of the splitting of the thymine dimer radical anion is asynchronously concerted or sequential [18]. Moreover, no theoretical evidence has so far been presented to confirm or refute the ultra-fast nature of the repair process, experimentally determined to be completed in 560 ps [18] in the case of the photolyase catalyzed mechanism. On the other hand, no experimental time-resolved data is to date available for the self-repairing photoreactivation mechanism.

Gas-phase quantum chemical studies predicted a stepwise mechanism in which the initial ring-opening process occurs spontaneously upon electron uptake [72, 74]. However, it was recently pointed out [98] that the cluster model underlying these gas-phase calculations is not sufficient to provide a realistic representation of the splitting process in condensed phase. Hydrogen bonding with surrounding water molecules or the complementary bases may stabilize the valence-bound state of the C4/C4' carbonyl groups where the electron is mainly localized and make it energetically more favorable than the dipole-bound state in the gas phase [99, 100]. Calculations including three water molecules showed that the hydrogen bond to the C4/C4' carbonyl groups has a dramatic effect on the reaction mechanism [73]. On the basis of these considerations Saettel et al. [73] suggested a quasi-concerted mechanism with an activation energy of 1.1 kcal/mol for the

4.3 Methods   45

cleavage of the C5-C5' bond and a subsequent barrierless breaking of the C6-C6' bond.

In this work, we have investigated the dynamics of the splitting of the thymine dimer radical anion in DNA, using a mixed quantum mechanical/molecular mechanics (QM/MM) method based on Born-Oppenheimer (BO) molecular dynamics. This study provides the first finite temperature dynamics of the reaction as opposed to the static theoretical pictures published so far, based on the minimum energy path studies along chosen reaction coordinates. Our approach exploits a density functional description for the thymine dimer along with a classical mechanics treatment of the molecular frame and the solvent, explicitly taking into account the steric and electrostatic effects of the environment. A set of 7 trajectories, initiated from an equilibrated run at constant pressure and temperature of the neutral TT structure, was used to sample the dynamics of the TT repair reaction. The statistical analysis of the performed trajectories shows that the mechanism behind the TT repair can be described as an asynchronous concerted mechanism, in which the breaking of the C5-C5' bond is spontaneous upon electron uptake and is subsequently followed by C6-C6' cleavage, providing an upper bound to the activation energy of the C6-C6' bond breaking of 2.5 kcal/mol.

4.3 Methods

4.3.1 Structural Model

The initial configuration was taken from the X-ray structure of a DNA decamer containing a cis-syn thymine dimer d[$GCTTAATTCG$]d[$CGAAT*T*AAGC$] (PDB entry code 1N4E, 2.5 Å resolution) [12]. The system was solvated with TIP3P water [101] in a rectangular box of 50 x 51 x 68 Å^3 and 18 potassium counter ions were added to neutralize its charge. The total system contained ~12'700 atoms (~4000 water molecules). The AMBER-parm99 force field [37] was adopted for the oligonucleotide and the potassium counterions. For the thymine dimer (T15,T16) the atom types at the C5/C5' and C6/C6' positions (see Fig. 4.1) were changed to "CT" atom type and at the H6/H6' positions to "H1" atom type. The modified charges for the lesion sites were computed using the RESP module of AMBER8 [102] for the isolated N-methyl derivatives of cis-syn TT following the same protocol as described in a previous paper [103]. A charge of 0.126 (the charge of the DNA backbone at the lesion sites) was imposed on the methyl groups and a RESP charge optimization was performed at the HF/6-31G* level of theory to yield a neutral model system. The RESP charges are provided in the appendix, together with the atom types and the additional force field parameters. Cross-links between T15/CT5-T16/CT5 and T15/CT6-T16/CT6 of the cis-syn dimer were also specified within the xleap module of AMBER8.

4.3.2 Classical MD Simulation

The system was treated within full periodic boundary conditions and electrostatic interactions were computed with the smooth Particle-Mesh Ewald (SPME) algorithm [104], using a cutoff of 10 Å for the real space part of the electrostatic interactions and the van der Waals term. A preliminary step involved the geometry optimization of the full structure using a conjugate gradient algorithm. Energy minimization was carried out in two steps: first with harmonic restraints on the DNA decamer, then without any restraint.

Subsequently, MD simulations were conducted with restrained DNA at constant volume (NVT ensemble), increasing the temperature from 0 to 300 K in five steps, for a total of 50 ps. The last restrained configuration at 300 K was then used for the unrestrained constant volume - constant temperature (NPT ensemble) simulations for 100 ps at T = 300 K. A MD run of $\approx$5 ns at constant pressure (1 atm) and temperature (300 K) was then performed to provide starting configurations for subsequent QM/MM calculations. All classical simulations were using an integration time step of 1.5 fs. The room-temperature simulations were achieved by coupling the system to a Berendsen thermostat [105]. The RMSD data reported in the appendix, confirm the stability of the unrestrained structure after full equilibration was achieved. All classical simulations have been performed with the AMBER suite of programs [102].

4.3.3 QM/MM MD Simulation

The QM/MM driver is based on the quantum mechanical program QUICKSTEP [42, 43] and the molecular mechanics driver FIST, which are both part of the freely available CP2K package [44]. The general QM/MM scheme recently de-

4.3 Methods

veloped for large scale molecular dynamics simulations [106, 56] is based on a multigrid technique for computing the electrostatic potential due to the MM atoms. The description of the quantum region is treated at the density functional theory (DFT) level. The QM region is made up of two thymine bases, named T15 and T16 according to the sequence of the PDB structure 1N4E, resulting in a total number of 30 QM atoms. The thymines have been cut at N1 and the valence of the terminal nitrogen atoms has been saturated by the addition of capping hydrogen atoms. The remaining part of the system, including water molecules and counter ions, has been modeled at the classical mechanics level with the AMBER-parm99 force field. A triple-ζ valence basis set with two sets of polarization functions, TZV2P [107], and an auxiliary plane-wave basis set with a density cutoff of 300 Ry were used to describe the wavefunction and the electronic density. It has already been shown that this kind of high-level basis set is necessary for accurate geometries as well as energetics [74]. Dual space pseudopotentials [108, 109] were used for describing core electrons and nuclei. We used the gradient corrected Becke exchange [47] and the Lee, Parr and Yang correlation functionals (BLYP) [48] and when adding an electron to the thymine dimer, the DFT calculations were performed within the local spin-density approximation (LSD). As DFT encounters difficulties in describing van der Waals interactions, dispersion corrected atom-centered potentials (DCACPs) [54, 110] have been used as a correction to the BLYP functional for all atoms, namely C, O, N and H. This correction allows to account for the van der Waals interactions between the two stacked thymines after splitting of the ring [111]. Energies were tested for convergence with respect to the wavefunction gradient ($5 \cdot 10^{-7}$ Hartree) and cell size, which was required to be no smaller than 14.8 Å to achieve a correct decoupling between the periodic images [112].

Among the 5 ns of the NPT run a set of 7 snapshots were randomly chosen and the geometries used as starting points for QM/MM simulations. First, the QM region of the QM/MM structures was relaxed by performing a geometry optimization, while the MM part was kept frozen. Then, a molecular dynamics run of the whole system was performed at 300 K using an NVT ensemble and collecting ≈ 2 ps of unconstrained QM/MM dynamics. Ab initio molecular dynamics (MD) simulations within the Born-Oppenheimer approximation were performed in the canonical (NVT) ensemble with an integration time step of 0.5 fs. The temperature was kept constant using a Nosé-Hoover thermostat [113, 114] with a time constant of 1 ps.

After the QM/MM equilibration run, an electron was added to the system and the simulation was continued for $\approx$ 2 ps, unless the breaking of both bonds was observed before.

QM/MM simulations are particularly sensitive to unphysical delocalization of the electronic density from the QM to the MM region, the so-called spill-out effect. An appropriately modified short-range Coulomb potential for the interaction between the QM and the MM parts was used to ensure that no spill-out occurs [115]. The spin density distribution of the radical anion thymine dimer indicates that the excess electron is fully localized on the QM part while being delocalized over both thymines (see Fig. 4.2).

Furthermore, since we are treating an odd-electron system, we assessed the influence of the self-interaction error (SIE) on the electronic structure by performing reference calculations with methods that are not suffering from SIE. The spin density was computed at the restricted open-shell Hartree-Fock (ROHF) and unrestricted MP2 (UMP2) levels. The spin density was also fully distributed on both thymines at these different levels of theory (data not shown) in line with our DFT results indicating that the SIE might be less important. Finally, the basis set superposition error is assumed to be negligible due to the extended basis set used.

The choice of including only the two thymines in the QM region relies on a recent study showing that proton transfer from a complementary adenine is an endothermic process and therefore unlikely in DNA [116]. Both complementary bases and water molecules around the two thymines have only interactions with the TT site through hydrogen bonds, which renders their description by a classical force field sufficient. A possible quantum description of the complementary bases or of the water molecules surrounding the two thymines would result only in a small numerical discrepancy on the energy barriers but it would not change the global picture of this work.

Nevertheless, the proton transfer is expected to take place in the active site of DNA photolyase [96] and may increase the repair quantum efficiency by stabi-

4.4 Results and Discussion

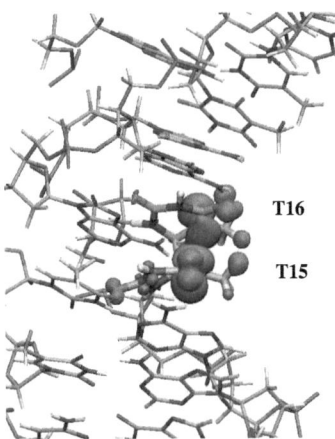

Figure 4.2: Spin density distribution of the optimized radical anion thymine dimer using a TZV2P basis set.

lizing the radical anion so that non-productive back-electron transfer is avoided and bond cleavage is favored. In the DNA environment the stabilization is not so crucial due to the small probability that the electron transfers back to an adjacent base. In fact, it was shown that electrons hop through DNA using pyrimidine bases as stepping stones [117].

4.4 Results and Discussion

7 different frames of the thymine dimer, namely CPD1-CPD7, are selected from the classical MD trajectory as starting points for 7 independent QM/MM MD simulations. The quality of the reference trajectory is checked in Table 4.1, where we report the RMSD values of the sampled structures compared to the X-ray structure. The main geometrical differences between the lesion site of the different conformations and the X-ray structure are the C5-C6 and C5'-C6'bonds which are essentially still π-bonds in the crystallographic structure, whereas upon cycloaddition they should have a complete σ-character. The low resolution (2.5 Å) of the X-ray data may be a possible explanation for this discrepancy.

In Table 4.2 we show the hydrogen bond distances between the lesion site and

	CPD1	CPD2	CPD3	CPD4	CPD5	CPD6	CPD7
X-ray struct.	0.69/2.62	0.57/2.84	0.62/2.65	0.78/2.81	0.76/2.94	0.73/2.84	0.74/2.90

Table 4.1: RMSD values [in Å] between selected snapshots along the classical trajectory and the X-ray structure. The first value refers to the cis-syn thymine dimer as shown in Figure 4.1, and the second value refers to the entire double helix.

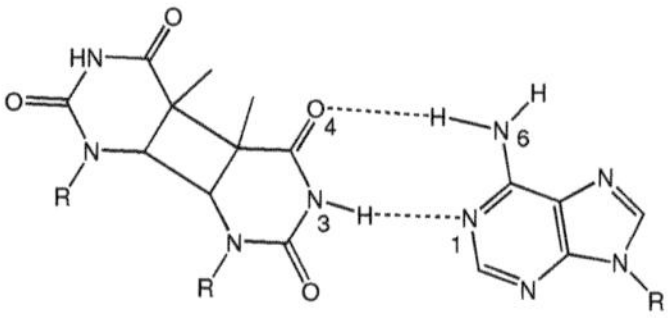

Figure 4.3: Hydrogen-bonding between the thymine dimer and a complementary adenine in DNA .

the complementary bases (see Fig. 4.3), where the comparison with respect to the X-ray data shows a lengthening of hydrogen bonds in our simulations.

Consistent with the crystallographic structure that finds the thymine T15 hydrogen bonds broken, the thymine T15 hydrogen bonds are weaker than those of thymine T16. Inspection of the trajectories of the neutral thymine dimer shows that the average bridging C5-C5' bond is longer than the C6-C6' bond (1.65±0.06 versus 1.61±0.05 Å, respectively), which is attributed to the repulsion between the two methyl groups. The cyclobutane puckering angle ∠C5-C6-C6'-C5' fluctuates around $14 \pm 4°$.

Electron Uptake. Upon electron uptake by the thymine dimer, the C5-C5' bond cleavage occurs spontaneously (see Figs. 4.5 and 4.6).

This initial ring-opening process can be rationalized as a delocalization of the singly occupied C4(C4')-O π^* orbital (SOMO) into the C5-C5' σ^* orbital [118]. The extra strain imposed on the puckered cyclobutane ring by the helical structure of DNA may enhance the rate of the C5-C5' cleavage by making the interaction of the SOMO with the cyclobutyl ring orbital larger [119]. Averaging the results of the 7 QM/MM simulations, we find that the C5-C5' distance fluctuates around 2.70 ± 0.14 Å and shows large fluctuations from 2.3 Å to 3.0

4.4 Results and Discussion

		CPD1	CPD2	CPD3	CPD4	CPD5	CPD6	CPD7	X-ray struct.
A5:H61	T16:O4	2.26	2.27	2.37	2.57	2.27	2.29	2.22	1.89
		(0.15)	(0.15)	(0.17)	(0.36)	(0.20)	(0.19)	(0.17)	(0.03)
A5:N1	T16:H3	1.90	1.96	1.95	2.04	1.95	1.96	1.98	1.85
		(0.12)	(0.13)	(0.16)	(0.26)	(0.16)	(0.12)	(0.13)	(0.03)
A6:H61	T15:O4	2.50	2.29	2.45	2.84	2.68	2.57	2.57	2.49
		(0.26)	(0.15)	(0.29)	(0.29)	(0.45)	(0.26)	(0.23)	(0.03)
A6:N1	T15:H3	1.98	2.12	2.02	2.00	1.99	2.06	2.05	2.20
		(0.12)	(0.31)	(0.14)	(0.17)	(0.12)	(0.21)	(0.18)	(0.03)

Table 4.2: QM/MM averaged lengths of hydrogen bonds (Å) for the A5-T16 and A6-T15 base pairs of the neutral cis-syn thymine dimer. Standard deviations are given in parenthesis.

Å. The C5-C6 and C5'-C6' bonds do still have σ-character as they display an average value of 1.54 ± 0.06 Å. The puckering angle rotates toward more positive values, with fluctuations up to $55°$ and an average of $35 \pm 4°$.

Spontaneous C6-C6' bond cleavage was observed in 4 out of 7 simulations. This bond cleaves within 500 fs after the C5-C5' bond breaking in all trajectories but one where the C6-C6' bond breaks 1900 fs afterwards (see Fig. 4.5). The C6-C6' distance increases strongly, fluctuating around 4.18 ± 0.47 Å while the C5-C5' distance rises at the same time and fluctuates around 3.88 ± 0.30 Å. These distances have to be compared with the values obtained by a 3 ns classical simulation on an undamaged decamer with the same base sequence, where C5-C5' and C6-C6' distances display an average value of 4.1 ± 0.3 and 4.5 ± 0.3, respectively. This indicates that the time scale of our QM/MM simulation (2 ps) does not allow to recover a fully undamaged DNA conformation.

Spin Density Distribution along the Reaction. The splitting of the cyclobutane ring can be monitored by spin density variations on relevant atoms as reported in Fig. 4.7. Upon ionization the electron is delocalized over both fragments and mainly found on the C5(C5') and O4(O4') atoms. During the breaking process a slight increase of the spin densities on C6 is observed. The

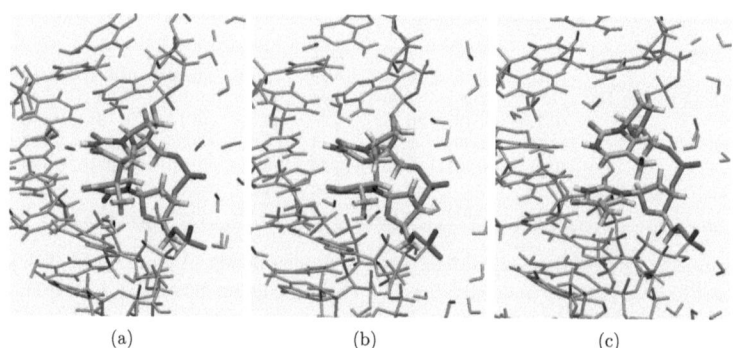

(a) (b) (c)

Figure 4.4: Representation of the thymine dimer splitting reaction during the simulation. The starting configuration is the thymine dimer (a). Upon electron uptake, the C5-C5' bond breaks spontaneously (b) and the C6-C6' bond breaks afterwards (c) according to an asynchronous concerted mechanism. The structure then relaxes to the canonical B-DNA conformation.

spin densities on O4 are close to a maximum and display a maximum on C5. However, on the other thymine, the spin densities on O4' and C5' are found close to zero. Thus, the electron is mostly localized on one thymine (T15) when the cleavage occurs with a maximum on C5 and an increase in the distribution of the spin density on C6. This favorable interaction between C5 and C6 allows for the formation of a double bond between them. After the C6-C6' bond dissociation the spin densities on O4 and C5 decrease significantly and the electron is mostly localized on C6 and C6'.

Hydrogen Bonding. Gas-phase quantum chemical studies at the MP2 [72] and B3LYP [74] levels of theory predicted stepwise splitting in which the C5-C5' bond breaks spontaneously upon electron uptake. Eriksson et al. [74] have calculated an activation energy of 2.3 kcal/mol for the C6-C6' bond cleavage which is the rate-determining step. It has been pointed out that such gas-phase calculations do not give an adequate representation of the reaction [73] since the valence-bound state of pyrimidines which hosts the extra electron can be stabilized by hydrogen bonding with water [75] or between base pairs [76]. Saettel et al. [73] suggested a quasi-concerted mechanism after including hydrogen bonding solvent molecules

4.4 Results and Discussion

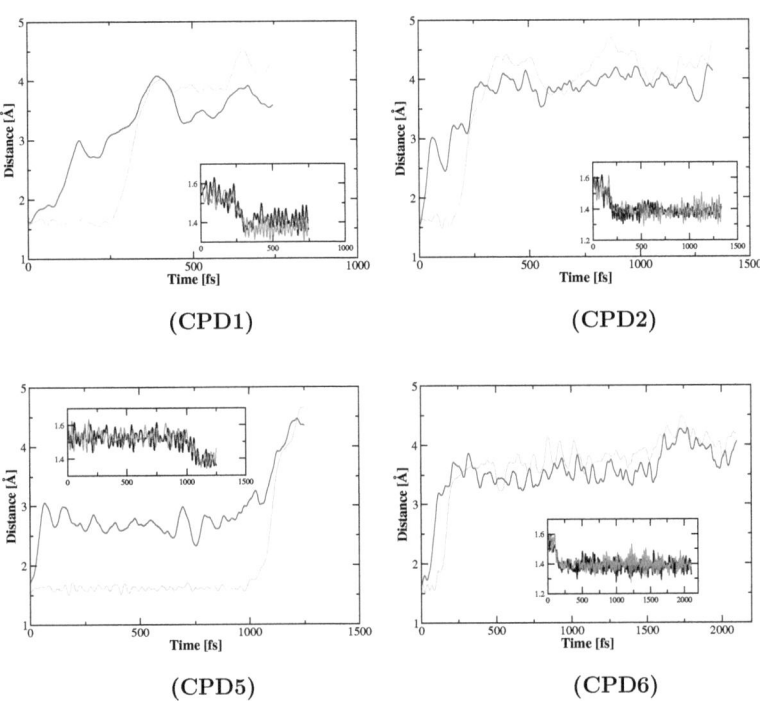

Figure 4.5: Time evolution of the C5-C5' (blue) and C6-C6' (green) bond lengths during the TT lesion reversion process along the 4 reactive trajectories. Inset: the C5-C6 (black) and C5'-C6' (red) bonds become double bonds when the C6-C6' bond is breaking.

in their model system. Through the introduction of three water molecules and single-point B3LYP/6-311++G^{**} calculations, the activation energy of the C5-C5' bond breaking is found to be 1.1 kcal/mol and the breaking of the C6-C6' bond occurs without a barrier. These considerations lead us to investigate the hydrogen bonds of each species with the complementary adenines (A) and water focusing our attention on the A:H61-T:O4 H-bonds which stabilize the developing negative charge on the C4 carbonyl group. In particular, in Figures 4.8 and 4.9, we show for both reactive and non-reactive trajectories the hydrogen bonds

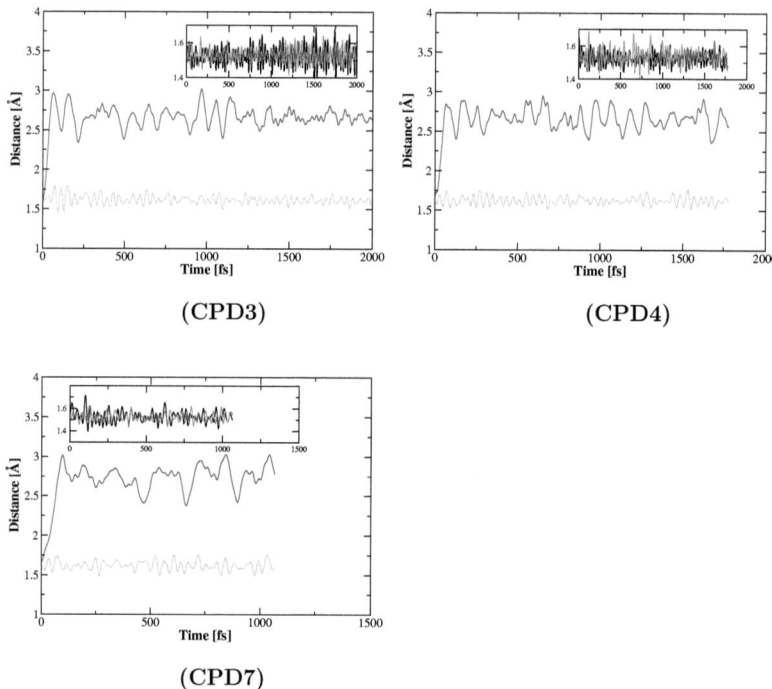

Figure 4.6: Time evolution of the C5-C5' (blue) and C6-C6' (green) bond lengths along the 3 non-reactive trajectories. Inset:the C5-C6 (black) and C5'-C6' (red) bonds.

involving one of the two thymines. It is evident that the hydrogen bond alone does not represent a driving variable for the reaction. On average the hydrogen bond involving the reactive species (see Fig. 4.8) is slightly shorter than the one from non-reactive species (see Fig. 4.9) but the almost similar values revisit the crucial role previously attributed to hydrogen-bonding [73]. O4 was also found to interact with a water molecule in the case of CPD1, CPD2 and CPD3 but no significant water interactions with the thymine dimer could be detected for the other species.

4.5 Bridging the time scale: Free Energy simulations

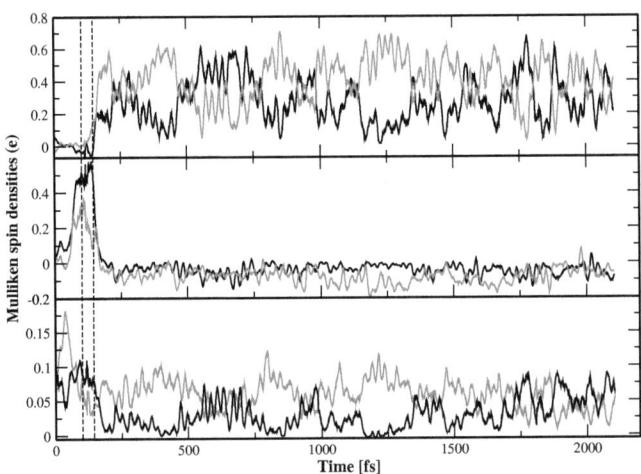

Figure 4.7: Time evolution of the Mulliken spin densities (e^-) of selected atoms shown only for CPD6 species. The breaking process of the C6-C6' bond occurs between the two dashed lines. Atoms of T15 are shown in black and atoms from T16 are shown in red. Upper panel: C6 (black) and C6' (red) spin densities. Middle panel: C5 (black) and C5' (red) spin densities. Lower panel: O4 (black) and O4' (red) spin densities.

4.5 Bridging the time scale: Free Energy simulations

The limited time scale accessible to our QM/MM MD simulations is in part responsible for the missing observation of the lesion repair. In other words, despite the minimal statistics, we have shown the evidence that a barrier is present in the self-repair mechanism and the lack of observing a spontaneous repair reaction in 3 out of 7 simulations is due to the problem of having trajectories trapped in local free energy minima.

To provide an estimate of the free energy barrier we have performed metady-

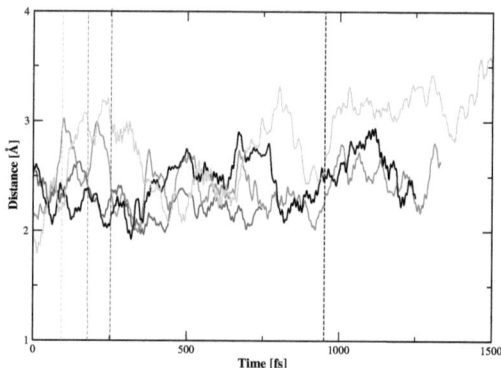

Figure 4.8: Variation with time of the A5:H61-T16:O4 hydrogen-bond length for the CPD1 species (blue), CPD2 species (red), CPD5 (black) and CPD6 species (green) where the breaking of the C6-C6' bond is observed. The dashed lines indicate when the breaking of the C6-C6' bond occurs.

namics runs [57, 120]. More details and references on this technique can be found in Section 2.4. It is worth noting that the choice of the collective variables (CV) in metadynamics is crucial for its successful application. In this case a natural choice was the C6-C6' bond length. Other collective variables would be necessary to fully characterize the real transition state in the free energy landscape. Since the aim of the current analysis is simply to provide an upper bound to the barrier we stress that the choice of the reaction coordinate is not only the most obvious but also the most reasonable in describing the repair process. The metadynamics runs were performed using Gaussian-shaped potential hills with a height of 0.3 kcal/mol and a width of 0.04 Å. The hills were spawned at intervals of 20 fs of QM/MM MD.

The free energy profiles of the three non reactive trajectories are plotted as a function of the CV in figure 4.10, and we can provide a value of 2.5 kcal/mol as a reasonable upper bound to the free energy barrier of the ring-splitting process.

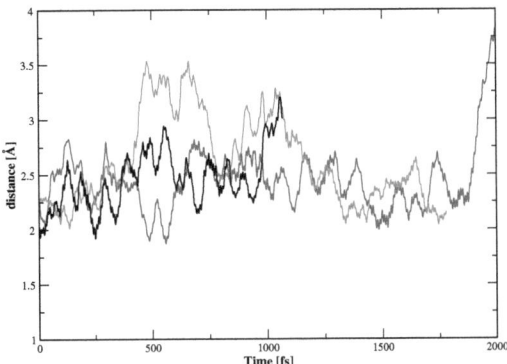

Figure 4.9: Variation with time of the A5:H61-T16:O4 hydrogen-bond length for the CPD3 species (blue), CPD4 species (red) and CPD7 species (black) where the breaking of the C6-C6' bond cannot be observed.

4.6 Conclusion

We have studied the photoactivated self-repair lesion mechanism of thymine dimer in DNA using a QM/MM scheme based on DFT for treating the quantum region. This work represents the first dynamical study of the repair process of the thymine dimer lesion in DNA and in general one of the few QM/MM studies that instead of relying on one simple MD trajectory performs several different statistically sampled runs.

Our computations, which take explicit account of the hydrogen-bonding network through the QM/MM interaction term, reveal an asynchronously concerted mechanism, providing an upper bound to the free energy barrier of 2.5 kcal/mol. The C5-C5' bond of the thymine dimer breaks spontaneously upon electron uptake and the subsequent C6-C6' cleavage occurs with a low barrier easily overcome with the thermal energy at room temperature. In one case the cleavage was also observed by performing a QM/MM simulation at 100K, confirming that the estimate of 2.5 kcal/mol is definitely an upper limit to the real free energy barrier. The exact determination of the energy barrier is beyond the goal of the present

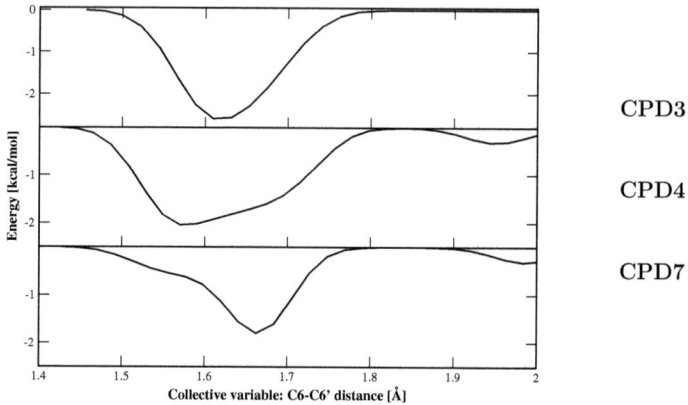

Figure 4.10: Free energy profiles of non-reactive trajectories (CPD3, CPD4, CPD7) with respect to the bond length C6-C6'.

investigation, but using novel methodologies [121, 122], it might be the subject for future studies.

The results confirm the experimental evidence [24] that a self-repair process for the thymine dimer is not only energetically feasible but also kinetically fast without any external catalytic protein. This provides an explanation for the observation of the photochemical stability of native DNA and definitely confirms the picture that, in a primordial world, life may have developed due to the intrinsic great resistance of the DNA/RNA molecules against the UV light.

4.7 Appendix

The modified classical force field for the thymine dimer, the full RMSD data of the equilibration processes as well as a detailed analysis of the 7 MD trajectories are provided in this appendix.

CPD1. The C6-C6' bond cleaves 250 fs after the C5-C5' bond breaking. Upon attachment of the electron A6:H61-T15:O4 and A5:H61-T16:O4 hydrogen bonds (H-bonds) fluctuate around 2.31 Å± 0.19 and 2.25 Å± 0.15, respectively. The A6:H61-T15:O4 H-bond shortly decreases before the C6-C6' cleavage reaching a minimum of 2.27 Å where the C6-C6' bond starts to increase. The A5:H61-T16:O4 H-bond has a similar value (2.30 Å). A6:N1-T15:H3 and A5:N1-T16:H3 H-bonds fluctuate both around 1.93 ± 0.10. One H-bond is found between T16:O4 and a water molecule with a minimum at 1.7 Å during the breaking process.

CPD2. Upon ionization, the C5-C5' bond breaks spontaneously followed by the cleavage of the C6-C6' bond after ∼ 120 fs. The A5:H61-T16:O4 H-bond shortly decreases before the C6-C6' cleavage reaching a minimum at 2.56 Å and the A6:H61-T15:O4 bond increases up to 2.70 Å when the cleavage occurs. After the splitting of the ring A6:H61-T15:O4 and A5:H61-T16:O4 H-bonds fluctuate around 2.77 Å± 0.38 and 2.48 Å± 0.23, respectively. One H-bond is found between T16:O4 and a water molecule with a value close of 2.00 Å during the cleavage.

CPD3. The cleavage of the C6-C6' bond cannot be observed during the simulation and the C5-C5' bond fluctuates around 2.66 Å ± 0.15. Upon attachment of the electron the A6:H61-T15:O4 and A5:H61-T16:O4 H-bonds fluctuate around 2.44 Å± 0.20 and 2.46 Å± 0.33, respectively. T16:O4 interacts with one water molecule and the bond fluctuates around 1.87 Å± 0.10.

CPD4. The cleavage of the C6-C6' bond cannot be observed during the simulation and the C5-C5' bond fluctuates around 2.67 Å ± 0.16. Upon attachment of the electron the A6:H61-T15:O4 and A5:H61-T16:O4 H-bonds fluctuate around

2.71 Å± 0.26 and 2.62 Å± 0.40, respectively. No H-bonds with water were found.

CPD5. Upon ionization, the C5-C5' bond breaks spontaneously followed by the cleavage of the C6-C6' bond after ∼ 950 fs. The A6:H61-T15:O4 H-bond and the A5:H61-T16:O4 H-bonds fluctuate around 2.47 Å± 0.17 and 2.40 Å± 0.21, respectively. No H-bonds with water were found.

CPD6. Upon ionization, the C5-C5' bond breaks spontaneously followed by the cleavage of the C6-C6' bond after ∼ 100 fs. The A6:H61-T15:O4 H-bond is decreasing shortly before the C6-C6' cleavage reaching a minimum of 2.10 Å . The A5:H61-T16:O4 H-bond has a similar value (2.18 Å). The A5:H61-T16:O4 H-bond then increases reaching a maximum of 4.32 Å after ∼ 1900 fs while the A6:H61-T15:O4 H-bond varies within the range of 1.87 to 3.06 Å . No H-bonds between the thymine dimer radical anion and water are observed before the cleavage of the C6-C6' bond.

CPD7. The cleavage of the C6-C6' bond cannot be observed during the simulation and the C5-C5' bond fluctuates around 2.70 Å ± 0.24. Upon attachment of the electron the A6:H61-T15:O4 and A5:H61-T16:O4 H-bonds fluctuate around 2.52 Å± 0.25 and 2.45 Å± 0.26, respectively. No H-bonds with water were found.

4.7 Appendix

Atom	Atom Types	RESP (e)
N1	$N\star$	-0.25
C2	C	0.60
O2	O	-0.57
N3	NA	-0.33
H3	H	0.31
C4	C	0.40
O4	O	-0.49
C5	CT	0.10
C5M	CT	-0.29
H51	HC	0.10
H52	HC	0.10
H53	HC	0.10
C6	CT	-0.01
H6	H1	0.09

Table 4.3: Atomic types and charges for T15 and T16 of cis-syn thymine dimer. The atom s are defined in the main text in Figure 1. RESP stands for restrained electrostatic potential derived charges obtained from Gaussian calculations [123].

Angle	K_θ (kcal $mol^{-1}\ rad^{-2}$)	θ_{eq}(deg)
CT-C-NA	70.0	115.38
CT-$N\star$-CT	70.0	129.84

Table 4.4: Additional force field parameters for thymine dimer which were taken direct ly from the X-ray coordinates. The force constants were taken from those of chemically similar groups.

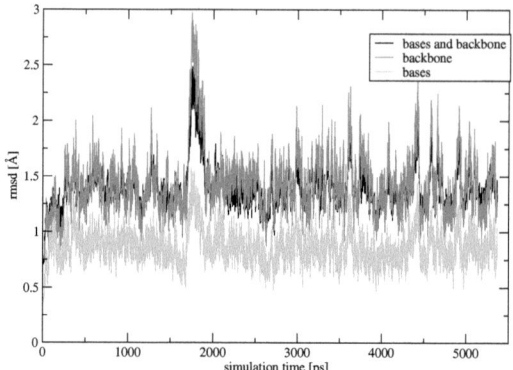

Figure 4.11: Root-mean-square deviations (rmsd) plotted as a function of simulated time for a DNA decamer containing a cis-syn thymine dimer.

Acknowledgment

The authors would like to thank Petra Munih for helpful discussion. Computer resources were provided by the Swiss National Supercomputing Center (CSCS) on IBM sp5 and by the University of Zurich on the Matterhorn Cluster.

Chapter 5

A QM/MM Investigation of Thymine Dimer Repair by DNA Photolyase

Abstract

DNA photolyase is a highly efficient light-driven enzyme which repairs the UV-induced cyclobutane pyrimidine dimer in damaged DNA. In this work, we investigate the repair reaction of the thymine dimer by means of hybrid quantum mechanical/molecular mechanical (QM/MM) dynamics simulations based on the X-ray structure of an enzyme-DNA complex. In analogy to the self-repair reaction, we find that the splitting mechanism of the cyclobutane ring is asynchronously concerted. Moreover, a few processes characterize the overall splitting mechanism: a continuous solvation reordering of the active site, a proton transfer from Glu283 to the thymine dimer and tight interactions between cationic side chains of Arg232 and Arg350 and the dimer. This suggests the important role of the active-site hydrogen-bond pattern in stabilizing the thymine dimer anion, leading to high repair quantum yields. Comparison of the repair efficiency with respect to the self-repair reaction is also discussed.

5.1 Introduction

Ultraviolet (UV) light, in particular UV-B radiation (290-320 nm), endangers all forms of life by the formation of genotoxic photoproducts of DNA. The most abundant damage caused to DNA is the cyclobutane pyrimidine dimer (CPD), formed between two adjacent pyrimidine nucleobases, mainly thymine, via a [2+2] photocycloaddition [90]. This lesion formation plays a crucial role in the initiation of UV-induced skin cancer [124]. Experimental [23, 24] and theoretical [125] investigations (see Chapter 4) have shown evidence for a self-repair mechanism in DNA under UV light exposure. However, the auto-catalytic process may not be an efficient channel for cells to repair the CPD lesion since it was shown that only the presence of readily oxidizable bases in the neighborhood of the lesion can boost the repair process. In order to convert the dimerized pyrimidines to their monomeric form, prokaryotes, plants and a variety of animals have developed a specific enzyme named DNA photolyase, which efficiently repairs CPD lesions. This enzyme is absent in placental mammals including humans which remove the CPD lesion through nucleotide excision repair. It was recently shown that the dimer repair in UVB-irradiated human skin can be enhanced by 45 % through topical application of DNA photolyase [124].

The mechanism of the catalytic cleavage of the cyclobutane ring relies on blue or near-UV light energy (300-500 nm) [95, 3, 96] and the overall process can be outlined as follows: the light energy, initially absorbed by an antenna pigment (8-hydroxy-5-deazaflavin HDF or methenyltetrahydrofolate MTHF), is transferred to a reduced flavin coenzyme ($FADH^-$) (Figure 5.1). The excited $FADH^-$ donates an electron to the CPD lesion, leading to a destabilization of the C5-C5' and C6-C6' bonds and thus to the splitting of the thymine dimer (TT) into the original bases. The ring-splitting reaction and the reduction of the $FADH^{\bullet}$ radical to $FADH^-$ by an electron back-transfer reaction from the repaired thymine monomers occur within 560 ps [18]. The high quantum yields of many DNA photolyases of up to 0.98 imply that non-productive electron back-transfer from the CPD lesion to the $FADH^{\bullet}$ radical before cleavage of the cyclobutane ring is completed is efficiently avoided by the enzyme. In this respect, the active-site solvation, the proximity of the adenine moiety of $FADH^-$ to the thymine residues as well as the substrate electric field are thought to play a crucial role in slowing down charge recombination [18, 126, 127].

The recently elucidated 1.8-Å X-ray structure of a complex between the *Anacystis nidulans* photolyase and a CPD-containing double-stranded DNA confirms that the enzyme flips the dimer out of the double helix into the highly polar active site cavity where it splits the cyclobutane ring [22] (Figure 5.2a).

Although many theoretical studies have attempted to elucidate the details of the repair mechanism, which are difficult to uncover by experimental means alone [71, 72, 74, 98], none of them analyzed the CPD repair reaction in the presence of the peculiar groups of the DNA photolyase active site. In Chapter 4, the self-repair process was investigated by QM/MM calculations of the TT within DNA and its water environment. We could show that in this case the mechanism of the splitting of the cyclobutane ring is asynchronously concerted, in which the breaking of the C5-C5' bond is spontaneous upon electron uptake and is subsequently followed by the C6-C6' bond cleavage [125]. Moreover, we addressed the ultra-fast nature of the self-repair process. Whether the repair proceeds sequentially or in a asynchronously concerted fashion is still an open question for the DNA photolyase-catalyzed reaction.

In this work, we have investigated the dynamics of the splitting of the thymine dimer radical anion within the DNA photolyase active site, using a mixed quantum mechanical/molecular mechanics (QM/MM) method based on Born-Oppenheimer (BO) molecular dynamics. To the best of our knowledge, this work provides the first modeling of the DNA photolyase reaction in the presence of the protein environment. The analysis is based on a set of 7 trajectories, initiated from an equilibrated run at constant pressure and temperature of the neutral TT structure and used to sample the dynamics of the TT repair reaction. Similar to our findings for the self-repair reaction, the statistical analysis of the performed trajectories also identifies the enzyme-catalyzed repair reaction as an asynchronous concerted mechanism, in which the breaking of the C5-C5' bond is spontaneous upon electron uptake and is subsequently followed by the C6-C6' cleavage, providing an upper bound to the activation energy of the C6-C6' bond breaking of 2.5 kcal/mol. Interestingly, in 5 out of 7 simulations we observed a proton transfer from Glu283 to the C4(T7) carbonyl oxygen of the thymine dimer radical anion (Figures 5.1 and 5.2b). The results are less decisive in determining whether or not the proton transfer plays a crucial role in the splitting mechanism. Whereas we observed a proton transfer in the non-reactive trajectory during a

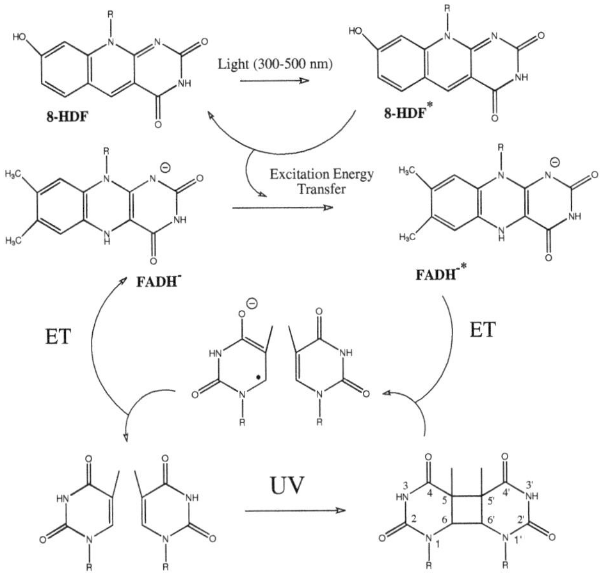

5.1 Introduction

Figure 5.1: Mechanism of repair of cyclobutane pyrimidine dimers (CPD) by DNA photolyase. 8-HDF: 8-hydroxy-5-deazaflavin, $FADH^-$: reduced and deprotonated flavin cofactor, ET: electron transfer. The atom numbering scheme is illustrated for the thymine dimer.

≈ 6 ps simulation, the repair reaction is also shown to proceed without proton transfer in two of the six reactive trajectories. In the latter case, the proximity of the cationic side chains of the neighboring arginines may play the crucial stabilizing role, which is usually provided by the proton transfer. This evidence corroborates a stabilizing role played by the proton transfer mechanism, though many other factors like the strain imposed on the dimer by DNA photolyase [70] and the dynamic active-site solvation [18] synergistically drive the ring-splitting reaction.

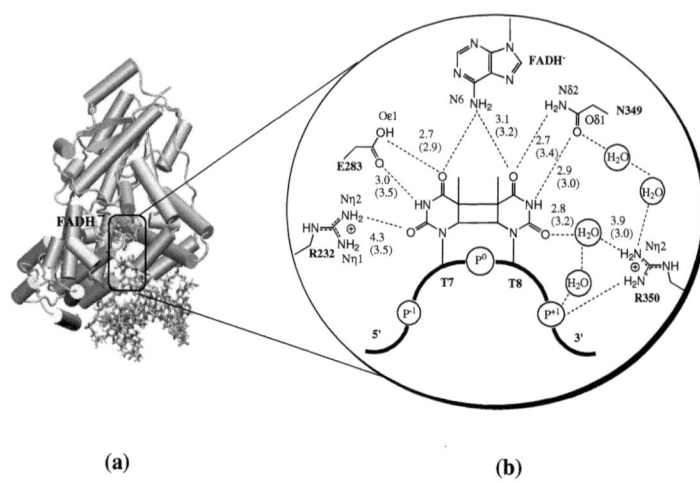

(a)   (b)

Figure 5.2: (a) DNA photolyase from *Anacystis nidulans* bound to double-stranded DNA with a CPD lesion (pdb code 1TEZ). (b) Characteristic interaction distances (Å) between the cis-syn thymine dimer and the active site in the classical optimized structure. For the sake of comparison, interaction distances (Å) revealed by the X-ray crystallography [22] are provided in brackets for the repaired thymine dinucleotide.

5.2 Methods

5.2.1 Structural Model

The initial configuration was taken from the X-ray structure of the *Anacystis nidulans* DNA photolyase bound to a DNA duplex containing a synthetic cis-syn thymine dimer lesion d[$ApTpCpGpGpCpT*pT*pCpGpC$]d[$CdGdAdAdGdCdC dGdA$] (PDB entry code 1TEZ, 1.8 Å resolution)(Figure 5.2 a) [22]. The synthetic intradimer formacetal linkage was replaced by a phosphate linkage to obtain a natural lesion. The 100 closest crystallographic water molecules around the complex were retained, preserving the water content of the active site. The crystallographic magnesium counter ion was also included in the model. The system was solvated with TIP3P water [101] in a rectangular box of 85 x 95 x 108 Å^3 and 18 potassium counter ions were added to neutralize its charge. The total size of the system is 72'500 atoms. The AMBER force field [37] was adopted

5.2 Methods

for the DNA photolyase, the oligonucleotide and the counter ions.

Although the crystal structure captures a repaired thymine dinucleotide, where the cleaved thymines have not yet been flipped back into the DNA helix, the interactions between the cleaved thymines and the active site are likely preserved before the splitting of the cyclobutane ring occurs (Figure 5.2 b). It is worth noting that the repaired CPD inside the active site is almost superimposable with a model of the intact lesion, indicating that the splitting of the CPD proceeds without major reorientations during the whole reaction course [96]. In contrast, our calculations are based on the cis-syn thymine dimer to simulate the repair reaction. The generation of the thymine dimer (T7,T8) and the computation of its RESP charges were described in Chapter 4. Gas-phase calculations were performed at the HF/6-31G* level of theory for the HDF and the reduced and oxidized FADH cofactors to compute RESP charges. The latter are provided in the appendix, together with the atom types and the additional force field parameters. Conventional (pH = 7) protonation states were chosen for the titrable residues. The histidine residues were assumed to be protonated on the ϵ-N atom except His132 which was protonated on the δ-N atom based on visual inspection of their putative H-bond patterns in the 1TEZ structure. Asp and Glu groups were taken as deprotonated except the active site Glu283 which is assumed to be neutral and to form a hydrogen-bond to the C4 carbonyl group of T7 [22, 96].

5.2.2 Classical MD Simulation

A 4-ns MD trajectory was performed at constant pressure (1 atm) and temperature (300 K) in order to provide starting configurations for subsequent QM/MM calculations. The classical MD simulation protocol was described in Section 4.3.2. All classical simulations have been performed with the AMBER suite of programs [102].

5.2.3 QM/MM MD Simulation

The QM/MM driver [106, 56] is based on the quantum mechanical program QUICKSTEP [42, 43] and the molecular mechanics driver FIST, which are both part of the freely available CP2K package [44]. The quantum part is treated at the density functional theory (DFT) level. This region is made up of two thymine bases, named T7 and T8 according to the sequence of the PDB structure 1TEZ, and the Glu283 side chain up to atom CB adding up to a total number of 40 atoms. The thymines have been cut at N1 (Figure 5.1) and the valence of the terminal nitrogen and CB atoms has been saturated by the addition of capping hydrogen atoms. The remaining part of the system, including water molecules and counter ions, has been modeled at the classical level with the AMBER force field, explicitly taking into account the steric and electrostatic effects of the DNA oligonucleotide, the enzyme and the solvent. A triple-ζ valence basis set with two sets of polarization functions, TZV2P [107], and an auxiliary plane-wave basis set with a density cutoff of 300 Ry were used to describe the wavefunction and the electronic density. It has already been shown that such a high-level basis set is necessary for accurate geometries as well as energetics [74]. Dual space pseudopotentials [108, 109] were used for describing core electrons and nuclei. We used the gradient corrected Becke exchange [47] and the Lee, Parr and Yang correlation functionals (BLYP) [48]. As DFT encounters difficulties in describing van der Waals interactions, dispersion corrected atom-centered potentials (DCACPs) [54, 110] have been used as a correction to the BLYP functional for all atoms, namely C, O, N and H. This correction allows to account for the van der Waals interactions between the two stacked thymines after splitting of the ring [111]. Energies were tested for convergence with respect to the wavefunction gradient ($5 \cdot 10^{-7}$ Hartree) and cell size, which was required to be no smaller than 20.0 Å to achieve a correct decoupling between the periodic images [112].

Among the 4 ns of the NPT run a set of 7 snapshots were extracted and the geometries used as starting points for QM/MM simulations. First, the QM region of the QM/MM structures was relaxed by performing a geometry optimization, while the MM part was kept frozen. Then, a molecular dynamics run of the whole system was performed at 300 K using an NVT ensemble and collecting $\approx$ 1 ps of unconstrained QM/MM dynamics. Ab initio molecular dynamics (MD) simulations within the Born-Oppenheimer approximation were performed in the canonical (NVT) ensemble with an integration time step of 0.5 fs. The temperature was kept constant using a Nosé-Hoover thermostat [113, 114].

After the QM/MM equilibration run, an electron was added to the system and the DFT calculations were performed within the local spin-density approximation (LSD), while using the RESP charges of the oxidized FADH.

5.3 Results and Discussion

QM/MM simulations with extended basis sets are sensitive to the unphysical delocalization of the electronic density from the QM to the MM region, the so-called spill-out effect. An appropriately modified short-range Coulomb potential for the interaction between the QM and the MM parts was used to ensure that no spill-out occurs [115]. The spin density distribution of the radical anion thymine dimer indicates that the excess electron is fully localized on the QM part while being delocalized over both thymines.

Furthermore, since we are treating an odd-electron system, we assessed the influence of the self-interaction error (SIE) on the electronic structure by performing reference calculations with methods that are not suffering from SIE. The spin density was computed at the restricted open-shell Hartree-Fock (ROHF) and unrestricted MP2 (UMP2) levels. The spin density was also fully distributed on both thymines at these different levels of theory (data not shown) in line with our DFT results indicating that the SIE might be less important. Finally, the basis set superposition error is assumed to be negligible due to the extended basis set used.

5.3 Results and Discussion

Classical Molecular Dynamics. Structural features. The enzyme-DNA complex was simulated for ≈ 4 ns at 300 K and atmospheric pressure. In Figure 5.3, the atom-positional root-mean-square-deviation (rmsd) of several parts of the complex with respect to the X-ray structure is shown. The oligonucleotide and DNA photolyase are both equilibrated after ≈ 1 ns. Although double-stranded DNA is bound to the DNA photolyase, it still retains some flexibility, as indicated by the rmsd of the backbone atoms (rmsd 1.81 ± 0.28 Å). The dynamics of the DNA atoms flanking the thymine dimer differs from that of the full oligonucleotide by a decrease in rmsd and a smaller fluctuation (rmsd 0.85 ± 0.17 Å). This value is obtained for the backbone atoms of the thymine dimer itself, the first adjacent base on the 5'-side of TT (cytosine) and the three adjacent bases on the 3'-side of TT (cytosine, guanine, cytosine) since these bases correspond to the region of enzyme binding along the DNA [128]. The reduced flexibility can be explained by the extensive manifold of salt bridges and hydrogen bonds (H-bonds) formed between the photolyase and the phosphates in the vicinity of the thymine dimer.

In particular, it was shown that the phosphates in the vicinity of the dimer interact extensively with the protein surface [128, 129]. Hence, these contacts at the photolyase-DNA interface play a crucial role in CPD-containing DNA recognition and in the stabilization of the flipped thymine dimer in the binding pocket.

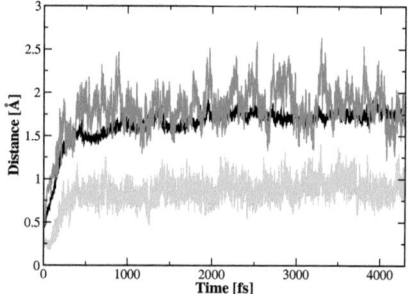

Figure 5.3: Root-mean-square deviations (rmsd) plotted as a function of simulated time for DNA photolyase (black), DNA backbone atoms (red) and DNA backbone atoms flanking the dimer (green).

Water molecules are continuously entering and leaving the active site in our simulations in agreement with a recent spectroscopic study which shows that the active-site solvation is a continuous dynamic process [18]. This peculiar characteristic was shown to control directly the catalytic reactions of DNA repair influencing the charge-separation between FAD and the CPD lesion, the efficiency of the ring-splitting and electron-return processes [18].

Superposition of MD averaged structures of active site moieties interacting with the CPD lesion shows that their position remains stable throughout the simulation except for the position of Arg232 (Figure 5.4). After releasing the restraints on the active site, Nη2(Arg232) significantly approaches Oδ1(Asn349), thereby drastically increasing the distance between Nη2(Arg232) and O2(T7) (from $\approx$ 3.5 Å to $\approx$ 6.5 Å). After $\approx$ 3500 ps of dynamics, the motion of the side chain of Arg232 culminates in a distinct flip-out, leading to the formation of a H-bond between Nη2(Arg232) and the phosphate of the stable flavin cofactor

5.3 Results and Discussion

(rmsd: 0.23 ± 0.03 Å) until the end of the simulation, whereas the distance between Nη1(Arg232) and O4(T7) gradually decreases (from ≈ 5.5 Å to ≈ 3.5 Å). The conserved Arg350 was found to be critical for substrate binding and discrimination by forming a salt-bridge with P^{+1} [129]. This interaction is maintained all along the 4-ns trajectory with an average value of 3.12 ± 0.74 Å (Table 5.7, appendix). The X-ray structure shows that the two thymines form H-bonds via their C4-carbonyl and N3-imide groups to the side chains of protonated Glu283, Asn349 and to the adenine N6 amino group of the flavin cofactor (Figures 5.1 and 5.2 b). These interactions are thought to be crucial to stabilize the radical anion CPD after electron transfer from $FADH^-$ so that non-productive back-transfer of the electron onto the $FADH^{\bullet}$ radical is avoided and bond cleavage is favored instead [18]. The hydrogen bonds with Asn349 and the adenine moiety are preserved during the whole duration of the simulation. The O4(T7)-Glu283 H-bond could be observed for ≈ 90 % of the time during the MD trajectory. Glu283 fluctuations randomly occur along the MD trajectory when its side chain rotates toward the carbonyl group of Val279, losing the H-bond with O4(T7). The flip-outs of the side chains of Arg232 and Glu283 which are well sampled within the ≈ 4-ns classical trajectory are unlikely to occur spontaneously during a first-principles QM/MM dynamics (≈ 10-ps time scale). On the basis of these observations, we selected from the classical MD simulations seven different configurations as starting points for subsequent QM/MM MD simulations. The frames CPD1-CPD3 and CPD4-CPD6 were captured before and after the rotation of the side chain of Arg232 toward the phosphate of the flavin cofactor, respectively. Furthermore, CPD7 corresponds to a configuration where Glu283 does not H-bond to the O4(T7) in order to assess the importance of this interaction in the repair process.

Mutagenesis studies indicate that the quantum yield for the splitting of the ring diminishes by 60 % in the E283A mutant [129]. Protonated Glu283 is thought to stabilize the thymine dimer radical anion after electron transfer from the flavin cofactor [22]. In fact, based on the X-ray structure, Essen et al. proposed a hypothetical repair mechanism that involves a proton transfer from protonated Glu283 to the C4 carbonyl oxygen of the 5' T base during the breaking process, leading to a stabilization of the radical anion [96].

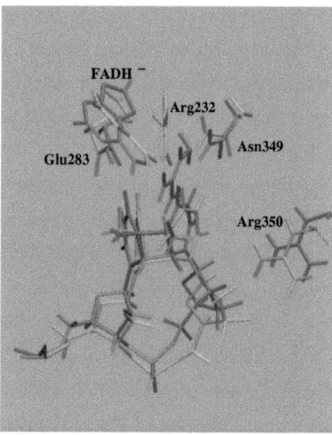

Figure 5.4: Superposition of average structures of the thymine dimer and active site moieties mentioned in the text. Before the flip of the side chain of Arg232 (0-3.5 ns) (red); after the flip of the side chain of Arg232 (3.5-4 ns) (green).

Repair Mechanism upon Electron Uptake. Initial QM/MM equilibrations ($\approx$ 1 ps) before adding the excess electron to the thymine dimer are performed to check the stability of the starting MD configurations. Inspection of the trajectories of the neutral thymine dimer shows that the average bridging C5-C5' bond is slightly longer than the C6-C6' bond (1.64 $\pm$ 0.05 and 1.60 $\pm$ 0.05 Å respectively), which is attributed to the repulsion between the two methyl groups. The puckering angle $\angle$C5-C6-C6'-C5' fluctuates around 13 $\pm$ 3°.

Upon electron uptake by the thymine dimer, the C5-C5' bond cleavage occurs spontaneously except for CPD5 whose C5-C5' bond cleavage is observed 400 fs after the electron transfer. This initial ring-opening process can be rationalized as a delocalization of the singly occupied C4-O4 π^* orbital (SOMO) into the C5-C5' σ^* orbital [118]. The free hydrated thymine dimer radical anions have been shown to cleave at a rate constant of $2 \times 10^6 s^{-1}$ [130, 131]. The rate of splitting the anionic ring of the thymine dimer in the active site is about 10^3-fold faster [18], indicating that the enzyme distorts the lesion in a manner that splitting of the ring is promoted [70]. In particular, the extra strain imposed on the puckered cyclobutane ring by the enzyme enhances the rate of the C5-C5' cleavage by making the overlap between the SOMO and the cyclobutyl ring orbital more

5.3 Results and Discussion

extensive [132, 119, 70, 133]. The torsion angle $\phi_{O4-C4-C5-C5'}$ describes the orientation of the C4=O4 double bond relative to the C5-C5' bond. In CPD5, $\phi_{O4-C4-C5-C5'}$ fluctuates around $-72 \pm 12°$ at the instant of the electron uptake and displays large oscillations from $-92°$ to $-54°$ ($-63 \pm 7°$) during the next 400 fs. In the case of the other frames, $\phi_{O4-C4-C5-C5'}$ fluctuates around more negative values ($-81 \pm 7°$) at the instant of the electron uptake, resulting in an increased orbital overlap. Thus, the weaker orbital-orbital overlap initially observed in CPD5 prevents immediate breaking of the C5-C5' bond. Averaging the results of the 7 QM/MM simulations, we find that the C5-C5' distance fluctuates around 2.68 ± 0.27 Å. The C5-C6 and C5'-C6' bonds still conserve a σ-character as they display an average value of 1.53 ± 0.04 Å . The puckering angle rotates toward more positive values, with fluctuations up to $40°$ and an average of $20 \pm 5°$.

C6-C6' bond cleavage was observed in 6 out of 7 simulations (Figure 5.5). The breaking process occurs shortly after C5-C5' bond cleavage (within 400 fs) in all trajectories but one where the C6-C6' bond breaks ca. 2800 fs afterwards. The C6-C6' distance increases strongly, fluctuating around 3.67 ± 0.31 Å while the C5-C5' distance rises at the same time and fluctuates around 3.53 ± 0.30 Å. Thus, our calculations predict an asynchronously concerted repair mechanism, in which the C5-C5' bond cleavage occurs spontaneously and is subsequently followed by the C6-C6' bond breaking process. In the case of CPD4, we could not observe the breaking of the C6-C6' bond within the QM/MM time scale. Therefore, we performed a metadynamics simulation [57, 120] to estimate an upper limit to the free energy barrier of the splitting reaction, as already described in Chapter 4. We obtained a value of 2.5 kcal/mol.

Comparison with Other Proposed Mechanisms. The results of this work propose an asynchronously concerted mechanism. We provide the first finite temperature dynamics of the repair reaction as opposed to previous static calculations [98]. Moreover, our model system includes the full enzyme environment, and in particular the manifold of interactions between the thymine dimer

Figure 5.5: Time evolution of the C5-C5' (blue) and C6-C6' (green) bond lengths during the TT lesion reversion process. The O4(T7)···H-Oε1(Glu283) distance is shown in red and the Oε1(Glu283)···H-Oε1(Glu283) distance is shown in black.

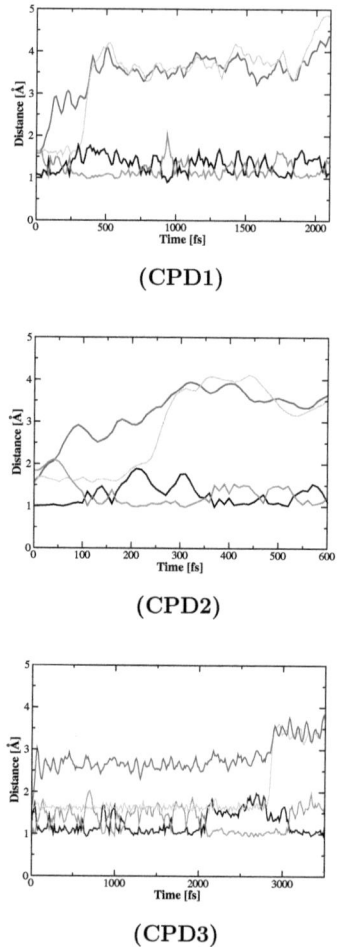

(CPD1)

(CPD2)

(CPD3)

and the active site. Gas-phase quantum chemical studies at the MP2 [72] and B3LYP [74] levels of theory predicted stepwise splitting in which the C5-C5' bond

5.3 Results and Discussion

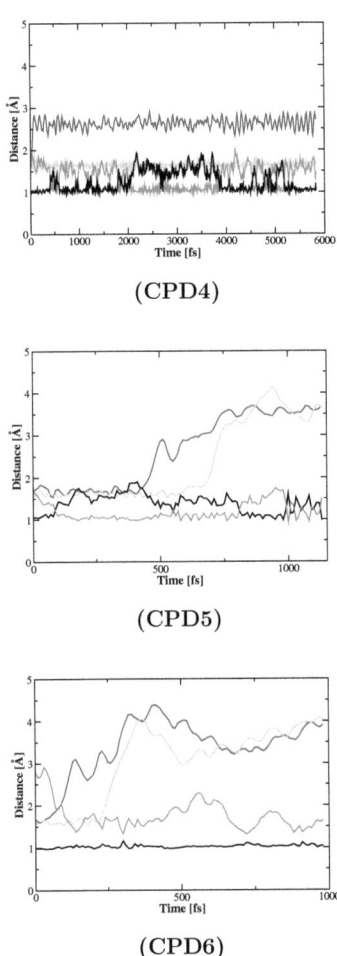

(CPD4)

(CPD5)

(CPD6)

breaks spontaneously upon electron uptake. Eriksson et al. [74] have calculated an activation energy of 2.3 kcal/mol for the C6-C6' bond cleavage which is the rate-determining step. However, it has been pointed out that such gas-phase calculations do not give an adequate representation of the reaction [73] since the

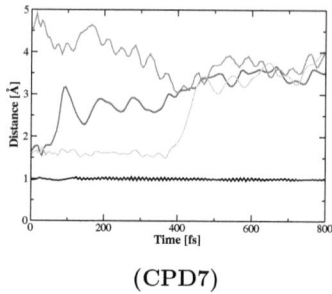

(CPD7)

valence-bound state of pyrimidines which bears the extra electron can be stabilized by hydrogen bonding with the active site residues or water molecules [75, 76]. Saettel et al. [73] suggested a quasi-concerted mechanism after including hydrogen bonding solvent molecules in their model system. Through the introduction of three water molecules and single-point B3LYP/6-311++G^{**} calculations, they predict the activation energy of the C5-C5' bond breaking to be 1.1 kcal/mol and the breaking of the C6-C6' bond to occur without barrier.

An asynchronously concerted mechanism has already been proposed in Chapter 4 for the self-repair reaction of a thymine dimer in double-stranded DNA. In the absence of the enzyme, the C6-C6' bond cleavage was observed in only 4 out of 7 simulations, and no crucial hydrogen-bonds have been identified. Interestingly, the stabilization of the dimer radical anion by hydrogen-bonding from active site moieties is a prerogative of the catalytic strategy of DNA photolyase so that bond cleavage is favored over non-productive back-transfer of the electron onto the $FADH^{\bullet}$ radical. In the case of the self-repair mechanism, the probability that the electron transfers back to an adjacent base is very low since it was shown that pyrimidine bases function as stepping stones along the excess electron transfer pathway though DNA [117]. However, the repair efficiency is compromised by the requirement of having readily oxidizable bases in the neighborhood of the lesion site. Moreover, the external energy supplied by the strain imposed by the polypeptide chain on the dimer may be more important than in the case of the DNA helix. In order to assess the influence of the mechanical strain on the splitting of the ring, we suggest to perform metadynamics simulations on both systems, i.e. (i) the DNA-enzyme complex and (ii) a DNA decamer containing a

5.3 Results and Discussion

thymine dimer, while switching off the charges of the classical part.

Role of Active Site Residues in the Reaction. These considerations lead us to analyze the dynamics of the interactions between the thymine dimer and the active site to identify the role of the conserved residues. Glu283, whose mutation to alanine impairs the enzyme activity by diminishing the quantum yield for the repair reaction by 60 % [129], is thought to stabilize the radical anion CPD by transferring a proton to O4(T7) [96]. Interestingly, our simulations show that a proton hops between Oε1(Glu283) and O4(T7) in 5 out of 7 trajectories. Moreover, the proton is always localized on O4(T7) during the C6-C6' bond cleavage process (Figure 5.5). In the case of CPD4, the proton is also found on O4(T7) during the C6-C6' bond breaking which is observed during a metadynamics simulation. After the splitting of the ring, the proton continues hopping between O4(T7) and Oε1(Glu283). Thus, the current hypothesis for a repair mechanism involving transient protonation of the CPD radical anion is confirmed by our calculations. In marked contrast, gas-phase quantum chemical studies predicted that a protonated CPD radical should experience a significant thermal barrier for cleavage [116]. Since we have observed the splitting of the thymine dimer with the proton localized on it, we argue that the absence of the barrier in the protein is essentially due to the stabilization of the active-site electrical field rather than a simple hopping mechanism. In case the hopping was a consistent factor in decreasing the thermal barrier, we would expect to observe the splitting of the dimer mostly when the proton is not localized on it. The only trajectory where no proton transfer to O4(T7) was observed are CPD6 and CPD7. In CPD6, the proton is always localized on Oε1(Glu283) and in CPD7 the H-Oε1(Glu283)-O4(T7) distance fluctuates around 3.85 ± 0.40 Å.

The observation that the ring splittings of CPD6 and CPD7 occur without protonation prompts us to also consider the roles of the positively charged side chains of Arg232 and Arg350. In the case of CPD6, no water-mediated saltbridges with the side chains of Arg232 and Arg350 can be observed. Instead, the two side chains are close enough to directly interact with O4(T7) and O2(T8) (H2-Nη1(Arg232)$\cdots$O4(T7) and H1-Nη2(Arg350)$\cdots$O2(T8) fluctuate around 2.49 ± 0.37 Å and 2.86 ± 0.44 Å, respectively) (Table 5.1). In CPD7, H-Nη1(Arg232) directly interacts with O2(T8) (2.63 ± 0.25 Å), displaying a minimal distance (2.1

Å) during the C6-C6' bond breaking process.

These observations suggest that the electrostatic contributions of the cationic side chains of Arg232 and Arg350 are sufficient to stabilize the dimer radical anion. Interestingly, alanine substitution at Arg350 also demonstrated a 60% decrease in quantum yield, indicating that Arg350 plays a key role in stabilizing the dimer [129]. In fact, it seems that a tight (water-mediated or direct) interaction between T8 and Arg232 or Arg350 is necessary to trigger the ring splitting as the electron is found to be localized on T8 when the cleavage of the C6-C6' bond occurs (see below). We observed that each O2 carbonyl group on T8 is tightly hydrogen-bonded either to water molecules or directly to the arginine side chains. The only exception being CPD4, whose O2 carbonyl group distances to water and arginine residues are above 3 Å most of the time during our simulations (Figure 5.6). This may be an explanation of the fact that we could not observe the C6-C6' bond breaking of CPD4 within the QM/MM time scale. Asn349 and the flavin cofactor appear to mediate the repair reaction by anchoring the dimer through hydrogen-bonds which are preserved during the entire process. The van der Waals interactions between the conserved tryptophans W286 and W392 and the thymine dinucleotide are maintained during the whole reaction course and these π-stacking effects may contribute to the stabilization of the dimer radical anion.

As already stated, the active-site solvation appears to be critical in order to delay the charge recombination by dynamically tuning the redox potentials of reaction species and stabilizing the charge-separated radical intermediates, leaving enough time to cleave the cyclobutane ring to reach a maximum-repair quantum yield [18]. However, the macroscopic details of its direct involvement in the ring-splitting reaction mechanism are not obvious. For example, in the non-reactive CPD4, hydrogen-bonds fluctuating around 1.9 Å between a water molecule and the CPD4 C4 carbonyl groups of T7 and T8 do not contribute to trigger the ring splitting. At the same time, the same kind of hydrogen bond is present in three other reactive trajectories, whereas in CPD6 the C6-C6' bond breaking occurs readily albeit no water molecules close to the thymine dimer could be detected. However, water-mediated bonds from O2(T8) to the cationic side chains of Arg232 and Arg350 may be crucial as already mentioned above.

5.3 Results and Discussion

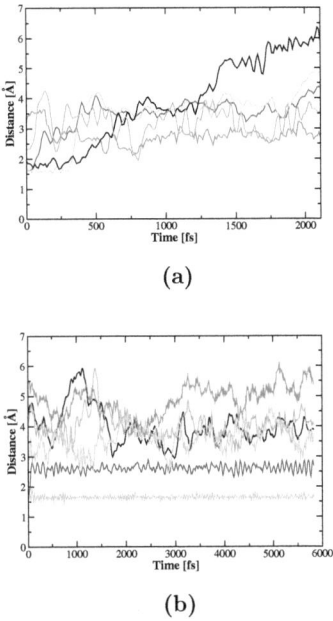

(a)

(b)

Figure 5.6: Time evolution of the interactions involving the O2 carbonyl group of T8 for a) CPD1 and b) CPD4, where the breaking process of the C6-C6' bond cannot be observed. The C5-C5' and C6-C6' bonds are indicated in blue and green, respectively. O2(T8)-WAT1 is in black, O2(T8)-WAT2 is in magenta, O2(T8)-H-Nη1(Arg232) is in red and O2(T8)-H-Nη2(Arg350) is in turquoise. In a), WAT1 leaves the active site after the breaking of the C6-C6' bond as a result of the dynamical solvation process. Arg350 is too far from the thymine dimer to have a direct interaction with O2(T8).

Spin Density Distribution along the Reaction. The splitting of the cyclobutane ring can be monitored by spin density variations on relevant atoms as reported in Figure 5.7 for the CPD2 trajectory. The excess electron is delocalized over both fragments and mainly found on the C5(C5') and O4(O4') atoms. It is interesting to notice the nearly perfect mirror symmetry of the spin densities on the cyclobutane ring carbon atoms during the breaking process. For instance, the spin densities show a progressive increase on C6' and a mirrorlike trend on

Distances	CPD1	CPD2	CPD3	CPD4	CPD5	CPD6	CPD7
H-Oε1@Glu283···O4@T7	1.21	1.30	1.36	1.33	1.18	1.76	3.85
	(0.19)	(0.30)	(0.27)	(0.26)	(0.21)	(0.34)	(0.40)
H-N6@FADH···O4@T7	3.59	2.84	2.81	3.18	3.04	4.43	3.53
	(0.71)	(0.34)	(0.37)	(0.34)	(0.30)	(0.72)	(0.64)
H-N6@FADH···O4@T8	2.64	2.69	2.83	3.41	2.45	2.23	2.80
	(0.25)	(0.44)	(0.38)	(0.57)	(0.25)	(0.13)	(0.35)
H-Nδ2@Asn349···O4@T8	2.41	2.22	2.45	2.41	2.23	2.39	2.39
	(0.32)	(0.18)	(0.29)	(0.21)	(0.17)	(0.23)	(0.23)
Oδ1@Asn349···H-N3@T8	2.75	2.87	3.53	4.27	2.58	2.64	2.95
	(0.30)	(0.19)	(0.43)	(0.58)	(0.24)	(0.17)	(0.41)
H1-Nη2@Arg350···O2@T8	3.82	4.13	4.12	3.59	3.36	2.86	3.52
	(0.37)	(0.27)	(0.50)	(0.41)	(0.41)	(0.44)	(0.50)
H-Nη2@Arg232···O2@T8	2.78	2.93	4.03	-	-	-	-
	(0.29)	(0.28)	(0.54)	-	-	-	-
H-Nη2@Arg232···O4@T7	3.65	5.08	4.44	-	-	-	3.86
	(0.78)	(0.43)	(0.28)	-	-	-	(0.51)
H2-Nη1@Arg232···O2@T8	-	-	-	4.83	4.50	4.69	2.63
	-	-	-	(0.48)	(0.38)	(0.43)	(0.25)
H2-Nη1@Arg232···O4@T7	-	-	-	4.81	4.36	2.49	-
	-	-	-	(0.51)	(0.41)	(0.37)	-
H-O@WAT···O4@T7	-	-	-	2.04	2.32	-	1.87
	-	-	-	(0.28)	(0.40)	-	(0.19)
H-O@WAT···O2@T8	3.87	2.06	2.40	3.99	3.59	-	-
	(1.42)	(0.17)	(0.69)	(0.61)	(0.59)	-	-

Table 5.1: QM/MM averaged lengths of selected active-site distances (Å) after electron transfer to the thymine dimer. Standard deviations are given in parenthesis.

C6. The spin densities on C5' display a peak with a maximum at 188 fs (0.73 e^-). A slight peak appears at the same time for the spin densities on C6 and C6'. Therefore, this peak may indicate precisely when a favorable interaction between C5' and C6' occurs leading to the formation of a double bond between them. The spin densities on O4' are increasing and a maximum is reached at 200 fs (0.20 e^-). However, on the other thymine, the spin densities on O4 and C5 are decreasing and are finally found close to zero. Thus, the electron is mostly localized on one thymine (T8) when the cleavage occurs with a maximum on C5' and an increase in the distribution of the spin density on C6'. After the C6-C6' bond dissociation the spin densities on O4(O4') and C5(C5') decrease significantly and the electron is mostly localized on C6 and C6'.

5.4 Conclusion

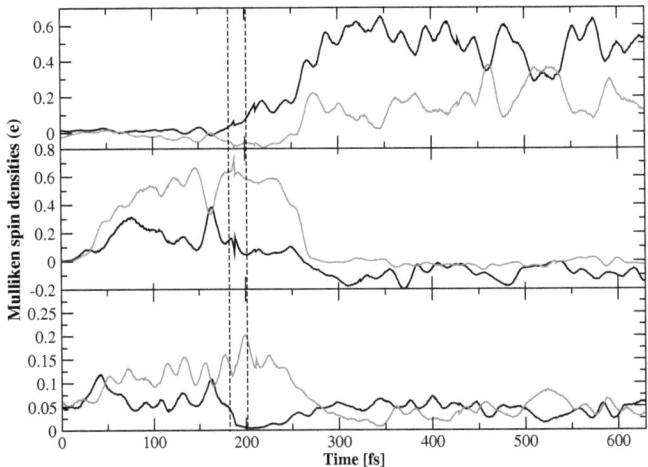

Figure 5.7: Time evolution of the Mulliken spin densities (e^-) of selected atoms showed for the CPD2 trajectory. The breaking process of the C6-C6' bond occurs between the two dashed lines. Atoms of T7 are shown in black and atoms from T8 are shown in red. Upper panel: C6 (black) and C6' (red) spin densities. Middle panel: C5 (black) and C5' (red) spin densities. Lower panel: O4 (black) and O4' (red) spin densities.

5.4 Conclusion

In this chapter, we have presented hybrid QM/MM simulations of the repair reaction of the thymine dimer by DNA photolyase. In contrast to the previous theoretical studies of the repair process, our model includes the protein environment and provides insight into the role of the different amino acids in the active site of the damaged DNA-enzyme complex. The statistical analysis of seven independent QM/MM trajectories identifies the reaction as asynchronously concerted, where C5-C5' bond cleavage spontaneously occurs after electron uptake and is subsequently followed by the C6-C6' bond breaking. C6-C6' bond cleavage could not be observed for only one trajectory. We thus applied the metadynamics approach for this species and obtained an upper bound to the free energy barrier of 2.5 kcal/mol. The repair mechanism involves a proton transfer from Glu283 to the T7 C4 carbonyl oxygen in 5 out of 7 trajectories. In the two other simulations, the crucial stabilizing role of Glu283 is taken over by the cationic side chains of neighboring arginines which are found very close to the thymine dimer. Tight (water-mediated or direct) interactions between the T8 O2 carbonyl oxygen and the side chains of Arg232 and Arg350 may play an important role in the splitting of the cyclobutane ring since such interactions were not detected for the species where the C6-C6' bond cleavage was not observed. Moreover, the electron is mostly found on thymine T8 during the C6-C6' bond breaking process.

Our study confirms (i) the picture of a highly-solvated active site in a continous dynamic exchange and (ii) the crucial stabilization by the adenine moiety of $FADH^{\bullet}$ and conserved residues to the radical anion dimer. Hence the very efficient repair of DNA photolyase can be explained by a sum of different synergistic effects, that can modulate the charge separation [18] and stabilize the radical anion, avoiding a back-transfer of the excess electron to the cofactor before dimer splitting.

5.5 Appendix

The modified classical force field and the atomic charges for the thymine dimer and the cofactors as well as selected active-site hydrogen-bond interactions from $\approx$ 4-ns MD simulations are provided in this appendix.

5.5 Appendix

Table 5.2: Atomic types and charges for T15 and T16 of cis-syn thymine dimer. The atoms are defined in the main text in Figure 5.1. RESP stands for restrained electrostatic potential derived charges obtained from Gaussian calculations [M.J. Frisch and al., *Gaussian 03*, Gaussian, Inc., Wallingford CT (2004)].

Atom	Atom Types	RESP (e)
N1	$N\star$	-0.25
C2	C	0.60
O2	O	-0.57
N3	NA	-0.33
H3	H	0.31
C4	C	0.40
O4	O	-0.49
C5	CT	0.10
C5M	CT	-0.29
H51	HC	0.10
H52	HC	0.10
H53	HC	0.10
C6	CT	-0.01
H6	H1	0.09

Table 5.3: Additional force field parameters for thymine dimer which were taken directly from the X-ray coordinates as in J. Antony, D. Medvedev, and A. Stuchebrukhov, J. Am. Chem. Soc. **122**, 1057 (2000). The force constants were taken from those of chemically similar groups.

Angle	K_θ (kcal mol^{-1} rad^{-2})	θ_{eq}(deg)
CT-C-NA	70.0	115.38
CT-$N\star$-CT	70.0	129.84

Table 5.4: Atomic types and charges for the FADH cofactor in the oxidized state $FADH^\bullet$ and reduced state $FADH^-$. The atoms are defined in Figure 5.8. RESP stands for restrained electrostatic potential derived charges obtained from Gaussian calculations [M.J. Frisch and al., *Gaussian 03*, Gaussian, Inc., Wallingford CT (2004)].

Atom	Atom Types	$FADH^\bullet$ RESP (e)	$FADH^-$ RESP (e)
P	P	1.099217	1.258925
O1P	O2	-0.790517	-0.833975
O2P	O2	-0.790517	-0.833975
O5*	OS	-0.388553	-0.470486
C5*	CT	-0.182041	0.057637
H51*	H1	0.160880	0.070039
H52*	H1	0.160880	0.070039
C4*	CT	0.020330	0.225847
H4	H1	0.083055	0.019979
O4*	OS	-0.361477	-0.494551
C3*	CT	0.144181	0.753861
H31*	H1	0.118991	-0.312995
O3*	OH	-0.472884	-1.111891
H32*	HO	0.320530	0.527244
C2*	CT	0.262419	0.349726
H21*	H1	1 0.053538	-0.026836
O2*	OH	-0.669096	-0.748776
H22*	HO	0.412712	0.425406
C1*	CT	0.026697	0.051654
H1*	H2	0.145222	0.148682
N9	N*	-0.077830	-0.014847
C8	CK	0.272038	0.318724
H8	H5	0.150084	0.110932
N7	NB	-0.639943	-0.662951
C5	CB	0.015382	0.015716

5.5 Appendix

Atom	Atom Types	$FADH^{\bullet}$ RESP (e)	$FADH^{-}$ RESP (e)
C6	CA	0.717186	0.757083
N6	N2	-0.913613	-0.929988
H61	H	0.399015	0.408272
H62	H	0.399015	0.408272
N1	NC	-0.814236	-0.814217
C2	CQ	0.534926	0.559004
H2	H5	0.068258	0.073913
N3	NC	-0.751250	-0.744795
C4	CB	0.413883	0.339088
N1A	N*	-0.297002	-0.294503
C2A	C	0.371543	0.482074
O2A	O	-0.677550	-0.633337
N3A	NA	-0.312027	-0.326398
H3A	H	0.276011	0.283262
C4A	C	0.481686	0.381864
O4A	O	-0.632272	-0.662242
C4B	CM	-0.041842	-0.051712
N5	N*	-0.384436	-0.382940
H5	H	0.312460	0.314449
C5A	CB	0.098229	0.091081
C6A	CA	-0.351999	-0.335582
H6A	HA	0.153648	0.158824
C7A	CA	0.095843	0.092149
C7M	CT	-0.166878	-0.193076
H7M1	HC	0.039350	0.050920
H7M2	HC	0.039350	0.050920
H7M3	HC	0.039350	0.050920
C8A	CA	0.061312	0.062686
C8M	CT	-0.319313	-0.281175
H8M1	HC	0.079436	0.072763

Atom	Atom Types	$FADH^{\bullet}$ RESP (e)	$FADH^-$ RESP (e)
H8M2	HC	0.079436	0.072763
H8M3	HC	0.079436	0.072763
C9	CA	-0.342988	-0.333138
H9	HA	0.209308	0.194219
C9A	CB	0.049190	0.033548
N10	N*	0.130265	0.153243
C10	CM	0.022462	0.030722
C12	CT	0.049106	0.010796
H121	H1	-0.018651	-0.005019
H122	H1	-0.018651	-0.005019
C13	CT	0.060495	-0.015670
H131	H1	0.108406	0.117953
O13	OH	-0.644738	-0.675449
H132	HO	0.373550	0.401517
C14	CT	0.131810	0.110524
H141	H1	0.066502	0.032395
O14	OH	-0.433667	-0.336270
H142	HO	0.283030	0.265969
C15	CT	0.153413	0.877178
H151	H1	0.056341	-0.201155
O15	OH	-0.642610	-0.708325
H152	HO	0.382064	0.390823
C16	CT	0.254578	-0.961460
H161	H1	0.003874	0.303101
H162	H1	0.003874	0.303101
O16	OS	-0.526948	0.020003
PF1	P	1.186795	0.158469
OF2	O2	-0.829660	-0.172121
OF3	O2	-0.829660	-0.172121
OF4	OS	-0.387743	0.155952

5.5 Appendix

Table 5.5: Atomic types and charges for HDF. The atoms are defined in Figure 5.9. RESP stands for restrained electrostatic potential derived charges obtained from Gaussian calculations [M.J. Frisch and al., *Gaussian 03*, Gaussian, Inc., Wallingford CT (2004)].

Atom	Atom Types	HDF RESP (e)
N1	N*	-0.451850
C1	C	0.608180
O1	O	-0.580220
N2	NA	-0.349190
C2	C	0.425080
O2	O	-0.544330
C3	CM	-0.009250
C4	CA	-0.106780
C5	CB	0.072760
C6	CA	-0.180790
C7	CA	-0.236250
C8	CA	0.389600
O3	OH	-0.559910
C9	CA	-0.289300
C10	CB	0.049330
N3	N*	-0.016680
C11	CM	0.253240
C12	CT	-0.000380
C13	CT	0.319100
O4	OH	-0.721640
C14	CT	-0.008810
O5	OH	-0.655260
C15	CT	0.107600
O6	OH	-0.651960
C16	CT	0.195170

Atom	Atom Types	HDF RESP (e)
O7	OH	-0.664560
H1	H	0.295130
H2	HA	0.173950
H3	HA	0.182830
H4	HA	0.170610
H5	HO	0.406800
H6	HA	0.194160
H7	H1	0.064270
H8	H1	0.064270
H9	H1	0.034280
H10	HO	0.428180
H11	H1	0.131650
H12	HO	0.495790
H13	H1	0.058040
H14	HO	0.456530
H15	H1	0.016820
H16	H1	0.016820
H17	HO	0.416970

Table 5.6: Additional force field parameters for the FADH and HDF cofactors which were taken directly from the X-ray coordinates as in J. Antony, D. Medvedev, and A. Stuchebrukhov, J. Am. Chem. Soc. **122**, 1057 (2000). The force constants were taken from those of chemically similar groups.

Angle	K_θ (kcal mol^{-1} rad^{-2})	θ_{eq}(deg)
$N\star$-CM-$N\star$	70.0	121.24
CA-CB-$N\star$	70.0	120.24
CB-$N\star$-CM	70.0	118.54
CA-CB-CA	63.0	119.75
CM-CA-CB	63.0	120.15
CM-CA-HA	50.0	119.99
C-CM-CA	63.0	117.24
C-CM-$N\star$	70.0	121.44

5.5 Appendix

Table 5.7: Selected active-site hydrogen-bond interactions from ≈ 4-ns MD simulations. Standard deviations are given in parenthesis.

H-bonds	MD averaged structure (Å)	X-ray structure (Å)
Oε1@Glu283···O4@T7	2.81(0.31)	2.86
Nδ2@Asn349···O4@T8	2.96(0.16)	3.27
Oδ1@Asn349···N3@T8	2.91(0.17)	2.97
Nη2@Arg232···O2@T7	6.07(0.92)	3.48
Nη2@Arg350···O2@T8	3.78(0.41)	4.07
Nη1@Arg350···O1@P^+1	3.12(0.74)	2.96
N6@FADH···O4@T8	3.02(0.22)	3.05

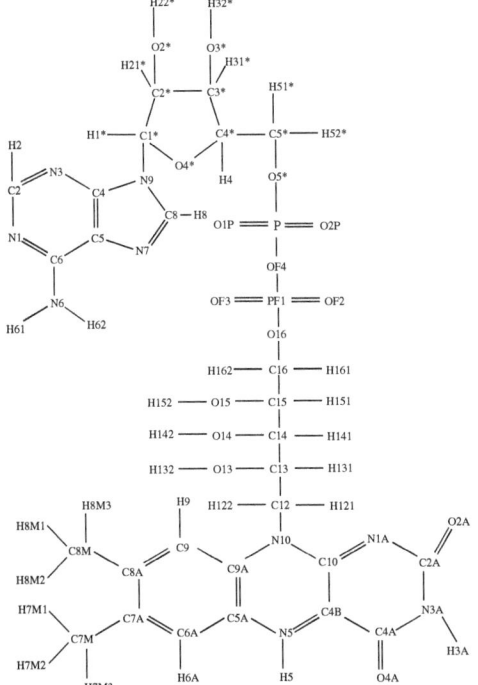

Figure 5.8: Atom labeling scheme of FADH showing the different components (from top to bottom): adenosine, phosphate, ribose and isoalloxazine.

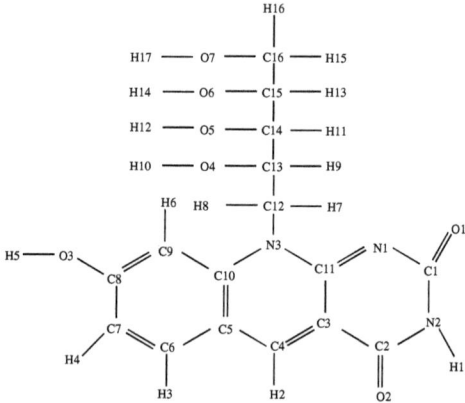

Figure 5.9: Atom labeling scheme of HDF.

Acknowledgment

Computer resources were provided by the Swiss National Supercomputing Center (CSCS) on IBM sp5 and by the University of Zurich on the Matterhorn Cluster.

Chapter 6

A Mixed QM/MM Metadynamics Study of Thymine Dimer Formation in DNA

Abstract

The singlet and triplet reaction pathways of the thymine dimer formation in a DNA decamer are explored by means of hybrid QM/MM metadynamics simulations. Precursors of the thymine dimer could be identified on both surfaces, but these conformations are not directly accessible from the ground-state surface and a significant barrier must be overcome on the excited state surfaces to reach them. Nonetheless, molecular dynamics simulations yield new insights into the relaxation pathways in the excited states leading to the thymine dimer. It is found that the triplet reaction proceeds over a diradical intermediate which decays to the thymine dimer via a crossing point with S_0. In contrast, the singlet mechanism does not involve any stable intermediate along the path towards a point of intersection with S_0.

6.1 Introduction

Upon absorption of ultraviolet (UV) light, DNA bases are promoted from their ground state to excited electronic states that can decay to mutagenic photoproducts [89]. The most abundant UV-induced damage caused to DNA is the cyclobutane pyrimidine dimer (CPD), formed between two adjacent pyrimidine nucleobases, mainly thymine (Figure 6.1) [90]. Dimerization proceeds through a [2 + 2] photocycloaddition reaction, in which the C5-C6 and C5'-C6' π-bonds of proximal pyrimidine bases are converted into two σ-bonds, resulting in the formation of a cyclobutane ring.

This lesion is efficiently repaired either by photoreactivation through exposure to blue or near-UV light (360-500 nm), or by nucleotide-excision repair pathways. In the photoreactivation process, the reversion of CPDs to base monomers is catalyzed by the enzyme DNA photolyase [3, 22]. Moreover, evidence for an autocatalytic process in which DNA repairs itself from UV-induced damage has been recently presented [23, 24, 125].

The photochemical yield of this lesion has been shown to depend on the deoxynucleotide (dN) sequence [92], the binding of sequence-specific proteins to DNA [94] and DNA conformation [93, 32]. In particular, it has been suggested that the spacing and the orientation of the pyrimidine nucleobases in their ground-state play a determinant role for the formation of the dimer [90]. DNA can encompass a wide variety of structural deformations and is highly dynamic [134]. This results in a vast number of conformations with local structural changes which continuously bring two adjacent pyrimidine bases into and out of favorable geometries for dimerization.

Data concerning the nature of the excited-state precursors to the pyrimidine dimer in DNA are sparse. UV absorption initially populates singlet excited $^1\pi\pi^*$ states, which overwhelmingly decay back to the electronic ground state via a process called internal conversion [135], revealing the intrinsic robustness of DNA against UV-light [89]. Low-lying triplet excited states can be populated either through intersystem crossing (ISC) along the ultrafast internal conversion pathway of the electronically excited singlet species [136, 137] or through photosensitization by an "external" or "endogenous" sensitizers via triplet-triplet energy transfer [90, 138, 139, 140].

Gas-phase theoretical studies have shown that the formation of CPDs can pro-

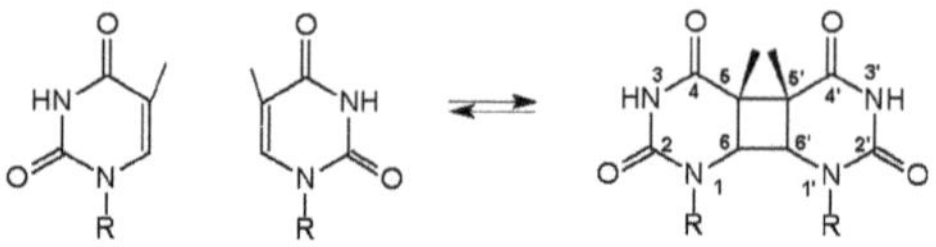

Figure 6.1: Formation and cycloreversion of the cis-syn thymine dimer.

ceed on both the triplet and singlet excited potential energy surfaces (PES) [85]. However, the photoreaction mechanisms completely differ from each other. While the triplet mechanism involves a diradical intermediate, subsequently leading to the dimer formation via singlet-triplet interaction, the singlet reaction occurs in a concerted fashion.

Though experimental studies have shown evidence for ISC to the lowest-lying triplet state of monomeric thymine [141, 142], its involvement in thymine dimer formation in polymeric DNA is still an unclear issue since a recent flash photolysis investigation could not detect it [143]. In contrast, thymine dimerization in DNA via triplet excited state can be induced by photosensitizers such as fluoroquinolones whose triplet energies lie as low as ≈ 63 kcal mol^{-1} [144].

Recent time-resolved measurements showed that the initial excited singlet state can decay to a dimer photoproduct in polymeric DNA if the nucleobases are properly oriented at the time of the excitation, clearly indicating the decisive role of the DNA conformation before light absorption [32]. For these particular conformations, thymine dimerization occurred on a femtosecond time scale, suggesting that the photoreaction proceeds along an essentially barrierless path connecting the initial $^1\pi\pi^*$ state to the ground-state photoproduct via a conical intersection. Some necessary ingredients for dimerization to occur have been proposed to characterize these reactive conformations. Among these geometrical requirements, base stacking, which reduces the distance between the two thymine bases, and the same conformational changes as in a dimer-containing double-stranded DNA are thought to be crucial criteria for reaction [32].

6.2 Methods   99

In this work, we explore both the singlet and the triplet reaction pathways of the thymine dimer formation in a DNA decamer by means of hybrid QM/MM simulations based on Born-Oppenheimer (BO) molecular dynamics at the GGA/DFT level. The free energy surfaces of the photoreaction were explored using metadynamics [57]. To the best of our knowledge, this work provides the first atomistic modeling of the thymine dimer formation in a DNA oligomer. The electronic structure of the first excited singlet state, S_1, is calculated within the restricted open-shell Kohn-Sham (ROKS) framework [64], which has been successfully applied to predict photochemical properties of excited molecules including nucleobases, in the gas phase and in condensed-phase systems [145, 146, 147, 148, 149, 150, 151].

6.2 Methods

6.2.1 Structural Model

The calculations were based on the X-ray structure of a DNA decamer containing a cis-syn thymine dimer d$[GCTTAATTCG]$d$[CGAAT*T*AAGC]$ (PDB entry code 1N4E, 2.5 Å resolution) [12]. The system was solvated with TIP3P water [101] in a rectangular box of 50 x 51 x 68 $Å^3$ and 18 potassium counterions were added to neutralize its charge. The total system contained ~12'700 atoms (~4000 water molecules). The AMBER-parm99 force field [37] was adopted for the oligonucleotide and the potassium counterions. The additional force field parameters, the modified atom types and the computation of the RESP charges of the thymine dimer (T15, T16) were described in Section 4.3.1. A 5-ns MD trajectory is performed at constant pressure (1 atm) and temperature (300 K). The classical MD simulation protocol was described in Section 4.3.2. All classical simulations have been performed with the AMBER suite of programs [102].

6.2.2 QM/MM MD Simulations

The QM/MM MD technique [55] we apply here has been developed as a part of an interface between the Car-Parrinello MD (CPMD) code [77] and the classical molecular dynamics package GROMOS96 [152]. The reactive part of the system, namely the two thymine bases T15 and T16, is treated at the quantum level

(DFT-BLYP[47, 48]) and the remaining part at the classical level (AMBER force field[37]). The QM/MM interface is modeled by the use of a link atom pseudopotential that saturates the electronic density of the QM region [110]. Two linking atoms were placed at C(1') where the two thymines have been cut. The electrostatic interaction between the QM system and the MM system are treated in a hierarchical scheme as described elsewhere [55]. An appropriately modified short-range Coulomb potential was used to ensure that no unphysical escape of the electronic density from the QM to the MM region occurs [55]. Bonded and van der Waals interactions between the QM and the MM region are treated with the standard AMBER force field [37]. Long-range electrostatic interactions between the MM atoms are taken into account by the P3M method [153]. The valence electrons are described by a basis set superposition error-free plane wave expansion up to an energy cutoff of 70 Ry. The interactions between valence electrons and ionic cores are described with norm-conserving Martins-Troullier pseudopotentials [78]. A 14.8 x 14.8 x 14.8 $Å^3$ supercell is used for the quantum part of the system. The inherent periodicity of the plane-wave calculations is circumvented solving Poisson's equation for non-periodic boundary conditions [79], while periodic boundary conditions are retained for the classical box. The local spin-density (LSD) approximation was used. As DFT encounters difficulties in describing van der Waals interactions, dispersion corrected atom-centered potentials (DCACPs) [54, 110] have been used as a correction to the BLYP functional for all atoms, namely C, O, N and H. This correction allows to account for dispersion interactions between the two stacked thymines [111]. Born-Oppenheimer molecular dynamics simulations are carried out with a time step of 0.24 fs. Constant temperature simulations are achieved by coupling the system with a Nosé-Hoover thermostat at 700 cm^{-1} frequency.

Three QM/MM simulations were performed using three different structural models. The first QM/MM simulation was initiated from the cis-syn dimer-containing DNA decamer, **TT-DNA**, by taking the last frame of the classical MD run, $TT - DNA_{cl}$. The second QM/MM simulation was started from a lesion-free decamer which retains the global conformation of a damaged decamer, **R-DNA**. Among the structural features required for a conformation to dimerize after electronic excitation, the same conformational changes as in a dimer-containing double-stranded DNA, such as the widening of both minor and

6.2 Methods

major groves and helix bending [12], are expected to be found [32]. **R-DNA** was prepared by performing a geometry optimization of $TT - DNA_{cl}$, using the steepest descent method (1000 steps) and the unmodified AMBER-parm99 force field [37]. The resulting rmsd's (all atoms) of the DNA moiety between the two structures is 0.19 Å. **R-DNA** conserved its global conformation along the QM/MM simulation, as indicated by the small rmsd between the QM/MM MD averaged equilibrated structure and $TT - DNA_{cl}$ (0.55 Å). Finally, the third QM/MM simulation was initiated from a canonical B-type DNA decamer with the same base sequence, **B-DNA**. **B-DNA** was generated using the nucgen module of the AMBER program package [102] and was classically equilibrated for ≈ 3 ns along the same protocol as for **TT-DNA**. The rmsd of the QM/MM MD averaged equilibrated structure relative to $TT - DNA_{cl}$ amounts to 5.65 Å.

The protocol of the QM/MM calculations includes an initial equilibration of the MD starting geometries: first the QM region of the QM/MM structures was relaxed by performing a geometry optimization, while the MM part was kept frozen. Then, the whole system is allowed to move and heat up to 300 K (≈ 1 ps of unconstrained QM/MM dynamics). This is followed by a 2-ps production run. Vertical excitation energies were calculated as average values over ≈ 30 snapshots taken at equal time intervals over the 2-ps trajectories of **R-DNA** and **B-DNA** at 300 K.

Singlet excited-state calculations are carried out with the restricted openshell Kohn-Sham (ROKS) approach [64, 154]. Triplet excited-state calculations are performed within the local spin-density approximation (LSD). Excited-state dynamics are run using the same settings as for the ground-state BO MD. For the sake of comparison, the vertical excitation energies to the singlet and triplet states were also computed using TDDFT [67] in the Tamm-Dancoff approximation [81, 155], which has been shown to yield excitation energies of valence states in good agreement with experimental values (mean absolute deviation less than 0.3 eV) [81].

6.2.3 Metadynamics

We adopt the metadynamics (MTD) methodology [57, 58] that allows the observation of "rare events" such as chemical reactions at moderate temperature. While the method is capable of providing accurate free energy information [63], in this study we also exploit its capacity to generate reaction paths and to explore probable configurations of the free energy surfaces in order to find properly oriented conformations for dimerization. More details and references on this technique can be found in Section 2.4.

The formation of the thymine dimer was studied by using the C5-C5' and C6-C6' distances as CVs. For the two CVs the mass M_α and the coupling constant are 20 and 0.4 a.u., respectively. Gaussian-type hills 3.137 kcal mol^{-1} high and approximately 0.15 Å wide were used to build up the V$(t,$s$)$. The hills were added every 50 MD steps, and velocities of the fictitious particles (CVs) were scaled to maintain a temperature of 300 K. A repulsive potential wall was placed at 5.6 Å for the first (C5-C5' distance) and second (C6-C6' distance) CVs in order to limit the distance between the two thymines.

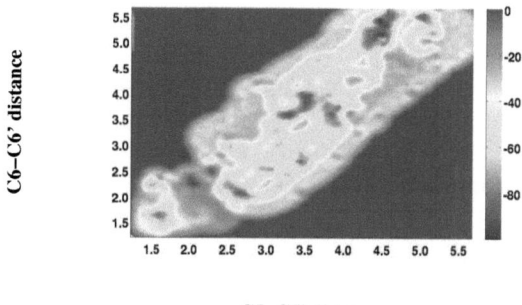

Figure 6.2: FES of the thymine dimer formation process in the ground-state as a function of the C5-C5' and C6 -C6' distances. The free energy is in kcal mol^{-1} and distances in Å.

6.3 Results and Discussion

6.3.1 Thymine Dimer Formation on Ground-State FES

Metadynamics is used to explore the ground-state free energy surface (FES) for the formation of the thymine dimer. It can be shown that this [2+2] cycloaddition reaction is forbidden by the rules of conservation of orbital symmetry for thermal reactions [70, 69]. Thus, a very high activation energy barrier is expected

6.3 Results and Discussion

for dimerization on the ground-state FES and the use of an efficient sampling technique such as metadynamics is necessary to observe the reaction process. Moreover, mapping of the FES allows us to examine in the next sections whether an excited-state configuration that can decay to the dimer corresponds to an accessible ground-state configuration.

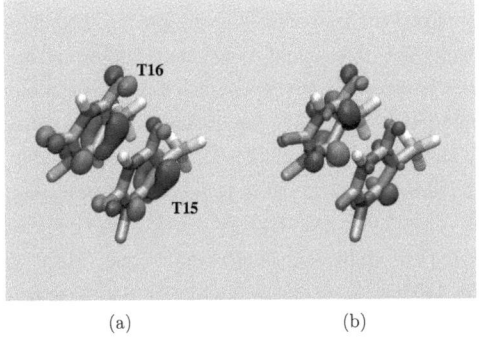

(a) (b)

Figure 6.3: HOMO (a) and LUMO (b) orbitals of the stacked thymine bases for a snapshot from the QM/MM MD simulation of **B-DNA** in S_0. The DNA decamer is not shown for the sake of clarity.

We first studied the back reaction by initiating the MTD simulation from the well of the thymine dimer. Starting from an equilibrated configuration of **TT-DNA**, the MTD trajectory first explores the basin of attraction of the thymine dimer until the accumulation of the MTD potential induces the simultaneous breaking of the C5-C5' and C6-C6' bonds, crossing a barrier of around 60 kcal mol^{-1} (Figure 6.2). Eriksson et al. have investigated the thermally induced [2 + 2] cycloreversion reaction in the gas-phase and have also found an activation energy barrier of ≈ 60 kcal mol^{-1} via a concerted mechanism [83]. This high activation energy barrier makes the thermally induced cycloreversion energetically inaccessible as expected for a "symmetry-forbidden" reaction.

The system then explores the broad region of the configurational space that corresponds to the conformations where the two thymines are not any more covalently linked. The explored structures are distributed along a valley where the two CVs increase simultaneoulsy with respect to each other. In order to accelerate the exploratory search of this broad region, the height of the hills was increased up to 10.0 kcal mol^{-1}. The use of lower hills would require a too long sampling time for this particular well. Even after 24 ps of metadynamics, the system did not cross back to the well of the thymine dimer. This could be explained by (i) the very high energy barrier and (ii) the not sufficiently long simulation time to span such a vast region. It is possible that, due to a high energy barrier between the vast basin of the separated thymines and the narrow basin of the dimer, the system has not reached the transition-state region yet. Therefore, further investigations with smaller hills (5 kcal mol^{-1}) to gain accuracy and a wall limiting even more the distance between the thymines (at 4.5-5 Å) as a compromise to reduce the sampling time are desirable. We can distinguish in the distance ranges d(C5-C5')=2.5-3.0 Å and d(C6-C6')= 2.0-2.5 Å a low-energy region where favorable configurations for dimerization could in principle be found since the two tymines are close to each other. QM/MM MD runs starting from a few selected structures show that they are not stable since both C5-C5' and C6-C6' distances increase up to $\approx$ 4.5 Å in 140 fs.

6.3.2 Vertical Excitation Energies of the Lowest Triplet- and Singlet-States

Figure 6.3 illustrates the Kohn-Sham highest occupied and lowest unoccupied orbitals for a snapshot corresponding to a stacked system (d(C5-C5')= 3.2 Å and d(C6-C6')= 3.0 Å) and shows that the HOMO and LUMO orbitals are distributed on both thymines. This observation is in line with recent gas-phase calculations on the π-stacked guanine-cytosine (GC) dimer by Varsano et al. [156]. As it can be seen from Figure 6.3, the charge density is primarily localized on the reactive C5(C5') and C6(C6') atoms with a predominant π-character for the HOMO and π^*-character for the LUMO.

Vertical excitation energies were computed from **R-DNA** and **B-DNA** in order to probe the direct influence of conformational changes. The global conformation of **R-DNA** is expected to bring the two thymines closer to each other. This is reflected by the C5-C5' and C6-C6' distances, which fluctuate around 3.7

6.3 Results and Discussion

± 0.2 Å and 4.0 ± 0.2 Å in **R-DNA**, whereas these distances are slightly longer in **B-DNA** with an average value of 3.8 ± 0.2 Å and 4.2 ± 0.2 Å, respectively (data not shown). It is worth noting that the C5-C5' and C6-C6' distances were fluctuating around larger values within the longer time scale of the 3-ns classical run of **B-DNA** (4.1 ± 0.3 Å and 4.5 ± 0.3 Å, respectively).

We first consider the lowest triplet state, T_1, of the two thymine bases. T_1 is dominated by a Kohn-Sham HOMO/LUMO transition. This transition is essentially a $\pi \rightarrow \pi^*$ excitation from the bonding π-orbital of the C5-C6 double bond to the anti-bonding π^*-orbital of the C5'-C6' double bond (Figure 6.3). For **R-DNA**, we obtain thermally averaged vertical excitation energies of 2.81 ± 0.16 eV and 2.82 ± 0.17 eV for the triplet state using TDDFT and LSD-DFT, respectively. Both methods yield a very similar shape for the T_1 surface. In the case of **B-DNA**, the computed vertical excitation energies are higher, 3.01 ± 0.11 eV and 3.03 ± 0.14 eV, using TDDFT and LSD-DFT, respectively. In agreement with our result, Eriksson et al. have reported a vertical excitation of 3.01 eV for the lowest triplet state of the two thymine bases at the TDDFT/B3LYP/6-31G(d,p) level using a dielectric constant of $\epsilon = 4.3$ to mimic the local surroundings of DNA [85]. Interestingly, it has been recently reported that compounds with triplet energies in the range 2.75-2.80 eV, such as fluoroquinolone derivatives, are able to photosensitize the thymine dimer formation, defining a threshold energy required for a given molecule to be capable of triggering the thymine dimer in DNA [144]. This threshold energy is in line with our excitation energies computed from **R-DNA**.

Second, we turn our attention to the lowest singlet state, S_1, of the two thymine bases. As for the triplet state, this excitation mostly involves the HOMO/LUMO orbitals for all the selected geometries along the QM/MM MD simulations of **B-DNA** and **R-DNA**. This excitation can thus be described as a $\pi \rightarrow \pi^*$ transition. For **R-DNA**, the thermally averaged vertical excitation energies amount to 3.12 ± 0.29 eV and 3.04 ± 0.19 eV, using TDDFT and ROKS, respectively. ROKS shows a systematic shift of all excitation energies to smaller values which has been discussed previously in the literature [64, 154]. Computation of vertical excitation energies from **B-DNA** yields slightly higher values, 3.28 ± 0.14 eV and 3.20 ± 0.14 eV, respectively. We note that similar excitation energies were computed for two thymine bases in gas-phase in Chapter 3 (3.20 eV at the TDDFT/PBE level). However, a computational study of two isolated thymine bases at the TDDFT/B3LYP/6-31G(d,p) level predicted a higher vertical excitation energy of 4.61 eV for the lowest excited singlet state [83].

In order to check our calculations, we have performed additional quantum-chemical calculations with the Turbomole program package [157, 158, 159], in which TDDFT is available both with gradient-corrected and hybrid density functionals. The calculations were carried out on the two isolated thymine bases taken from the same snapshots using the TZVP basis set. Test calculations with the larger basis set aug-cc-pVTZ do not yield significantly different energies. The B3LYP functional is used for testing the performance of a hybrid functional. Hybrid functionals are known to often give an improved description of excitation energies compared to pure density functionals [158].

Use of the BLYP functional gives excitation energies of 3.37 ± 0.11 eV for **B-DNA** and 3.26 ± 0.15 eV for **R-DNA** for the $S_0 \rightarrow S_1$ transition. These energies are in excellent agreement with the values obtained from gas-phase calculations performed at the same geometries with CPMD/BLYP/TDDFT (mean absolute deviation less than 0.04 eV). Thus, the energies of the system in gas-phase are slightly blueshifted with respect to the energies obtained in DNA.

Using B3LYP, the S_1 state is also mainly dominated by a HOMO/LUMO excitation, but the energies are higher compared to the BLYP values. We obtained 4.23 ± 0.15 eV for **B-DNA** and 4.03 ± 0.20 eV for **R-DNA**. Both values are within the UV region (200-320 nm) known to induce the fomation of the thymine dimer. Triplet excitation energies display a blueshift of 0.13 eV, which is about seven times lower than the shift for the singlet excited state: 3.14 ± 0.10 eV for **B-DNA** and 2.94 ± 0.12 eV for **R-DNA**. Thus, using BLYP yields too low excitation energies, but we believe that the description of the excited-state FES is not affected since this redshift is systematic and that we could show in Chapter 3 that the ordering of excited states is not altered.

In conclusion, the smaller average distance between the thymine bases in **R-DNA** leads to a redshift of about 0.2 eV with respect to **B-DNA** for both $S_0 \rightarrow T_1$ and $S_0 \rightarrow S_1$ transitions ($\pi \rightarrow \pi^*$). TDDFT and LSD-DFT yield very

6.3 Results and Discussion

similar values for the triplet excitation energies. Moreover, our computed values are in fair agreement with experiment [144] and a previous theoretical study [85]. In the case of the singlet excitation energies, it is obvious that BLYP gives too low values. Test calculations show that the singlet excitation energies are better described with B3LYP since they lie in the UV region, in agreement with experimental data. Triplet excitation energies are much less affected by the use of B3LYP. The use of B3LYP along with plane wave-based calculations is prohibitively expensive and we will employ pure density functionals here.

6.3.3 Thymine Dimer Formation on Excited-State FES

Thymine Dimer Formation on Triplet FES

Triplet excited-state MD simulations are performed within the local spin-density approximation (LSD). This method yields vertical excitation energies which are very similar to the ones calculated using TDDFT (see above). A ≈3-ps QM/MM MD simulation of the two thymine bases in DNA is carried out in T_1 starting from an equilibrated ground-state configuration of **R-DNA** (Figure 6.4).

The most significant geometrical change observed during the dynamics in T_1 compared to the ground-state dynamics is the conversion of the C5-C6 π bond to a σ bond, in agreement with previous gas-phase studies [140, 160]. The C5-C6 and C5'-C6' bonds fluctuate around 1.37 ± 0.02 Å and 1.49 ± 0.03 Å respectively, showing that the triplet-state spin density is localized on T16, while T15 displays ground-state structural features (Figure 6.5a and Figure 6.4, inset). Moreover, T16 undergoes a considerable out-of-plane distortion with the dihedral angle $\phi_{H6-C6-C5-C7}$ increasing from $2 \pm 5°$ in S_0 to $10 \pm 22°$ in T_1. Mulliken spin densities show mostly a localization of charge on the C5' and C6' atoms (0.591 and 0.657 e^-) as for the isolated triplet thymine [85]. We observe an increase of both C5-C5' and C6-C6' distances without any sign of dimerization. This clearly indicates that the lesion formation in T_1 cannot be observed within the QM/MM time scale (Figure 6.4).

Therefore, a metadynamics simulation is performed at 300 K based on a frame selected along the ground-state QM/MM MD dynamics of **R-DNA**, ST_T (d(C5-

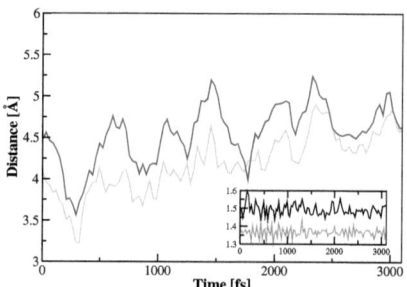

Figure 6.4: Time evolution of the C5-C5' bond length (green) and the C6-C6' bond length (blue) along one QM/MM MD simulation in T_1. Inset: the C5-C6 bond length (red) and the C5'-C6' bond length (black). The triplet excitation is localized on the T16, resulting in a longer C5'-C6' bond length.

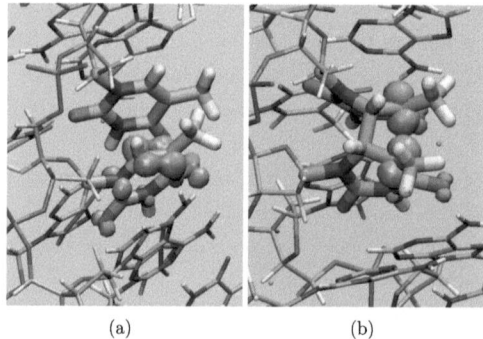

(a)   (b)

Figure 6.5: Spin density for the starting ST_T and intermediate geometries IT_T of the thymine bases along the metadynamics simulations in the T_1 excited state.

C5')= 3.57 Å and d(C6-C6')= 3.93 Å.

1. $ST_T \to IT_T$

As the initial well is progressively filled with the hills, the system crosses a barrier of around 17 kcal mol^{-1} to a region in the phase space where we can distinguish local minima characterized by shorter C6-C6' distances, finally leading to the

6.3 Results and Discussion

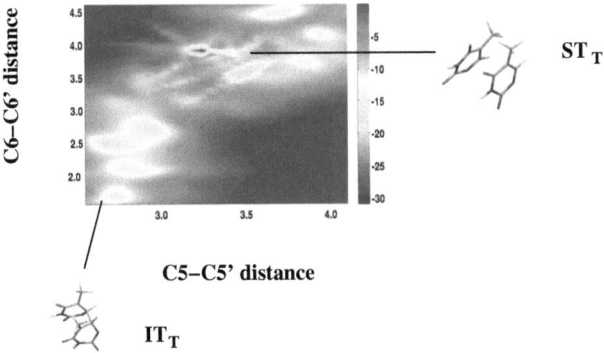

Figure 6.6: FES of the thymine dimer formation process in the T_1 excited state as a function of the C5-C5' and C6-C6' distances. The free energy is in kcal mol^{-1} and distances in Å. The starting ST_T and intermediate configurations IT_T are also shown.

formation of the C6-C6' bond (Figure 6.6). A QM/MM MD run at 300 K starting from the structure which has the shortest C6-C6' distance, IT_T (d(C5-C5')= 2.65 Å and d(C6-C6')= 1.58 Å), indicates that this intermediate is stable over the entire ≈3-ps simulation period in T_1. IT_T is a diradical species where the spin densities are mainly localized on the C5 and C5' atoms (0.710 and 0.768 e^-) (Figure 6.5b). Thus, a spin transfer from T16 to T15 occurs through the reaction as shown in Figure 6.5. The SOMO1 is mostly of σ character in which the larger atomic contributions arise mainly from the subunit containing the C5(C5')-C6(C6') bonds, while the SOMO2 is predominantly of σ^* character with the largest contribution from the C5-C5' bond (Figure 6.7). After reaching IT_T, the backward transition to the well corresponding to the two non-covalently linked thymines proceeds over a similar energy barrier. Eriksson et al. have optimized the lowest-lying triplet stationary points for the cycloaddition reaction of two thymine bases at the TDDFT/B3LYP/6-31G(d,p) level [85]. Their 0K quantum chemical calculations predicted a smaller activation energy barrier (6.8 kcal mol^{-1}) for the crossing towards the diradical intermediate with the C6-C6' bond formed. The discrepancy between the values of the barriers can be accounted for by the fact that this study does not take into account the effects of the surrounding DNA and the solvent. The constraints imposed by the DNA double-strand may be responsible for the higher barrier.

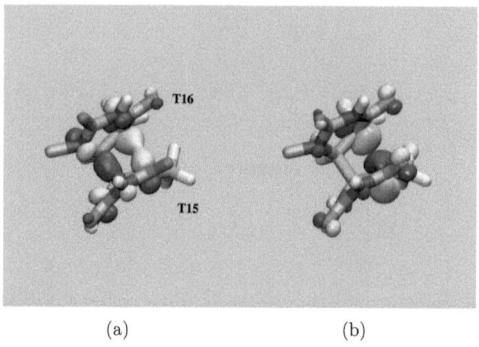

Figure 6.7: SOMO1 (a) and SOMO2 (b) at IT_T in the T_1 excited state. The surrounding DNA is not shown for the sake of clarity.

2. $IT_T \rightarrow TT$

Eriksson et al. suggested that the thymine dimer cannot be formed on the triplet-state surface only since the formation of the dimer from the diradical intermediate requires a high energy barrier (58.5 kcal mol^{-1} at the TDDFT/B3LYP/6-31G(d,p) level) [85]. The ring closure is rather expected to occur through singlet-triplet interaction after formation of the diradical intermediate. We checked this hypothesis by first computing the vertical excitation energy for the $S_0 \rightarrow T_1$ transition at the IT_T geometry. This transition is of pure HOMO/LUMO character and TDDFT calculations yield a small negative value (-7.22 kcal mol^{-1}), indicating that IT_T lies in a surface crossing region between S_0 and T_1. Calculations within LSD give a small negative value as well (-8.76 kcal mol^{-1}). In contrast, Eriksson et al. found an energy gap between the diradicals and the singlet-triplet crossing points ranging from 14 to 15 kcal mol^{-1}. Then, we started a standard BO QM/MM ground-state simulation from IT_T. The LSD approximation was used since the starting diradical system represents an open-shell ground state. As shown in Figure 6.8, the C5-C5' distance very quickly decreases, leading to the thymine dimer already after $\approx$ 70 fs. Thus, we conclude that the diradical intermediate IT_T can be indeed considered as a precursor of the thymine dimer.

6.3 Results and Discussion

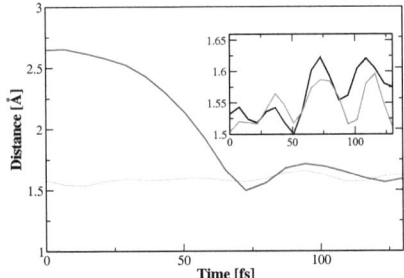

Figure 6.8: Time evolution of the C5-C5' bond length (blue) and the C6-C6' bond length (green) along a QM/MM ground-state simulation starting from IT_T and resulting in the thymine dimer formation. Inset: the C5-C6 bond length (red) and the C5'-C6' bond length (black) are both σ-bonds, fluctuating around 1.54 ± 0.03 Å and 1.56 ± 0.04 Å, respectively.

The question arises as to whether IT_T can be directly reached by vertical excitation from an accessible ground-state conformation. From the plot of the FES in S_0 (Figure 6.2), we can see that the ground-state conformation corresponding to IT_T (d(C5-C5')= 2.65 Å and d(C6-C6')= 1.58 Å) does not belong to probable configurations due to its too short C6-C6' bond length. In order to investigate the possibility to access such a reactive diradical species on the triplet surface, we selected one configuration from the triplet metadynamics simulation with a longer C6-C6' bond length, namely IT_1 (d(C5-C5')= 2.58 Å and d(C6-C6')= 1.81 Å). Once in the ground-state, IT_1 relaxes rapidly towards the basin of attraction of the separated thymines. In case the system remains on the triplet excited-state surface, the C5-C5' and C6-C6' distances of this stable C6-C6' cross-linked intermediate fluctuate around 2.91 ±0.15 Å and 1.69 ±0.05 Å, respectively (Figure 6.9, upper pannel). One encountered configuration with a relatively short C6-C6' distance (d(C5-C5')= 2.85 Å and d(C6-C6')= 1.65 Å) was put back to the ground-state and the relaxation in S_0 led to the thymine dimer (Figure 6.9, lower pannel). Thus, a barrierless route on T_1 towards a T_1/S_0 crossing point from where the system evolves to the thymine dimer could be found. It is worth noting that the T_1/S_0 crossing points revealed by our simulations involve a diradical intermediate on the triplet surface. However, this surface crossing region is not thermally accessible on a picosecond time scale from non-covalently linked thymine bases (free-energy barrier ≈ 17 kcal mol^{-1}) and does not correspond to any accessible ground-state conformation.

We have also performed ground-state simulations starting from triplet configurations with longer C6-C6' bond lengths (d(C6-C6') > 1.75 Å and d(C5-C5') ≤ 3.0 Å), but they resulted in an increase of both C5-C5' and C6-C6' distances and failed to form the dimer. Finally, ground-state simulations from selected structures in the basin of the two separated thymines were carried out as well, but no dimer formation could be observed.

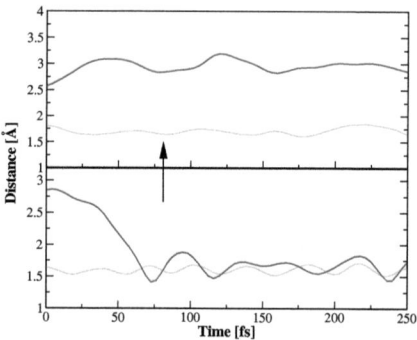

Figure 6.9: Time evolution of the C5-C5' bond length (blue) and the C6-C6' bond length (green) starting from IT_1 in T_1 (upper panel). One conformation (indicated by an arrow) along the simulation in T_1 is relaxed in S_0 (lower pannel), resulting in the thymine dimer formation.

Thymine Dimer Formation on Singlet FES

Singlet excited-state MD simulations are carried out within the ROKS scheme. This methodology has been shown to describe excited states in condensed-phase systems properly despite some conceptual shortcomings, while vertical excitation energies are red-shifted with respect to the experimental and TDDFT results (see above) [145, 146, 149, 154]. In addition, we expect the ROKS forces to be very

6.3 Results and Discussion

similar to the ones computed within TDDFT since the nature of the transition is almost of a pure HOMO/LUMO type. A $\approx$3-ps QM/MM MD simulation of the two thymine bases in the DNA decamer is carried out in S_1 starting from an equilibrated ground-state configuration of **R-DNA**. The C5-C6 and C5'-C6' bonds fluctuate around 1.38 $\pm$0.03 Å and 1.48 $\pm$0.03 Å respectively. Thus, the C5'-C6' bond becomes a single bond upon excitation and the singlet-state spin density is localized on T16. As for the triplet excited-state, T16 undergoes a significant out-of-plane distortion with the dihedral angle $\phi_{H6-C6-C5-C7}$ increasing from 2$\pm$5° in S_0 to 17$\pm$25° in S_1. The C5-C5' and C6-C6' bonds fluctuate around 4.03 $\pm$0.15 Å and 4.60 $\pm$0.17 Å respectively, showing no sign of dimerization.

We have also performed a QM/MM MD simulation in S_1 starting from a ground-state equilibrated configuration of **TT-DNA**. In this case, the C6-C6' bond breaks very quickly, indicating that the thymine dimer corresponds to a very high energy region on the S_1 surface. During the simulated time of 1.5 ps, the C5-C5' and C6-C6' bond lengths fluctuate around 1.75 $\pm$0.10 Å and 2.87 $\pm$0.21 Å, respectively. Ten frames were selected along the S_1 dynamics and were subsequently relaxed in S_0 to check whether this stable C5-C5' cross-linked intermediate is a potential precursor of the thymine dimer, but once in the ground-state the system was always attracted towards the basin corresponding to the two separated thymines. Since it was not possible to observe the dimer formation within the time scale of a QM/MM simulation, we switched to the MTD approach as for the triplet-state dynamics.

The starting point was one frame from the QM/MM MD simulation in S_1 initiated from **TT-DNA**, namely IT_{S1} (C5-C5'= 1.69 Å and C6-C6'= 2.76 Å). Figure 6.10 shows the FES projected onto the plane of the 2 CVs. The system first crosses over a barrier of $\approx$ 25 kcal mol^{-1} to the wide valley where the two thymines are not any more covalently bound. The height of the hills was increased up to 10.0 kcal mol^{-1} as for the ground-state metadynamics simulation. Then, the system recrosses a barrier of $\approx$ 60 kcal mol^{-1} back to the well where the C5-C5' bond is formed. At IT_{S2} (C5-C5'= 1.47 Å and C6-C6'= 2.61 Å), the species with the shortest C5-C5' bond length, the vertical excitation energies amount to 16.42 kcal mol^{-1} with TDDFT and 6.33 kcal mol^{-1} with ROKS. This species is unstable since a short QM/MM MD run in S_1 shows that both C5-C5' and C6-C6' bonds immediately increase. When released to the ground-state, IT_{S2} evolves

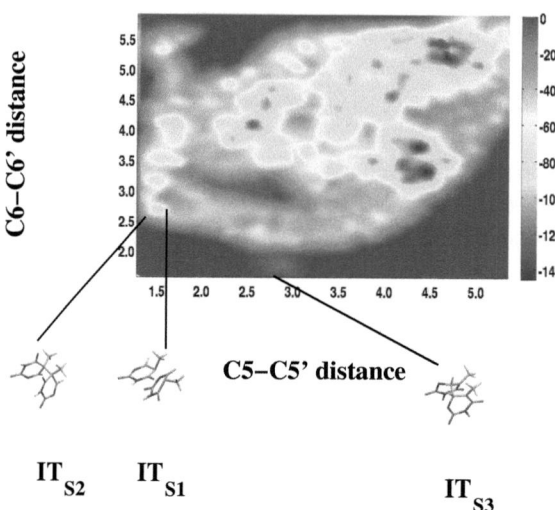

Figure 6.10: FES of the thymine dimer formation process in the S_1 excited state as a function of the C5-C5' and C6-C6' distances. The free energy is in kcal mol^{-1} and distances in Å. The species IT_{S1}, IT_{S2}, IT_{S3} are also shown.

towards the thymine dimer in $\approx$ 50 fs. However, IT_{S2} does not correspond to an accessible ground-state conformation (see Figure 6.2). A few other precursors of the dimer with slightly longer C5-C5' bond lengths can be found in this well along the metadynamics simulation. However, these unstable species are out of reach since they do not correspond to probable ground-state conformations and are separated by a very high barrier from the configurational space of the two non-covalently bound thymines in S_1. This very high activation energy barrier can be explained by the formation of C5-C5' cross-linked structures in which the methyl groups on C5 and C5' present unfavorable interactions. Next, the system returns to the basin of the two separated thymines. A few structures were put back to the ground-state but none of them could lead to the thymine dimer. After spending a long time exploring this broad region of the configurational space, the sysyem proceeds along a pathway where the C6-C6' distance lies below 3 Å and the C5-C5' distance decreases from $\approx$ 4.5 Å to $\approx$ 1.5 Å. The configuration which has the shortest C6-C6' distance, IT_{S3} (C5-C5'= 2.76 Å and C6-C6'= 1.63 Å), is

6.3 Results and Discussion

tested as a potential precursor of the thymine dimer. First, a QM/MM simulation is carried out in S_1 to check whether IT_{S3} is a stable structure. Both C5-C5' and C6-C6' bond lengths immediately increase, showing that IT_{S3} does not lie in a minimum in S_1. Then, a ground-state QM/MM MD run is performed within LSD and the formation of the dimer occurs in less than 50 fs. However, IT_{S3} does not correspond to an accessible conformation in the ground-state due to its too short C6-C6' bond length. Moreover, IT_{S3} does not lie close to a minimum of the S_1 FES and a significant barrier must be overcome to reach IT_{S3} from the basin of non-covalently bound thymines.

In order to gain insight into the dimerization reaction mechanism, a structure with a longer C6-C6' bond length, namely IT_{S4} (C5-C5'= 2.64 Å and C6-C6'= 1.86 Å), is selected from the metadynamics simulation. This configuration is very unlikely to be found on the ground-state FES (see Figure 6.2) and when released to S_0, an immediate increase of both C5-C5' and C6-C6' bond lengths is observed, preventing the formation of the dimer. A QM/MM MD run is performed in S_1 and the C6-C6' bond length decreases up to 1.59 Å before breaking after $\approx$ 200 fs (Figure 6.11, upper pannel). IT_{S4} is thus not a stable species. At d(C6-C6')= 1.59 Å, the vertical excitation energies amount to 11.30 kcal mol^{-1} with TDDFT and 7.02 kcal mol^{-1} with ROKS. This configuration is put back to the ground-state and the system evolves towards the thymine dimer (Figure 6.11, lower pannel). Thus, a barrierless route on S_1 towards a point of intersection with S_0, which is characterized by a small C6-C6' distance and leads to the thymine dimer, could be found. This rapid decay mechanism can be described as concerted since no intermediate could be detected on this pathway. This prediction can explain why the dimerization is a ultrafast process when a reactive ground-state conformation is excited. It is also in line with previous static ab initio calculations of the two isolated thymines which predicted a concerted pathway for dimerization in S_1 [83]. We also selected a singlet configuration with a longer C6-C6' bond length (C5-C5'= 2.65 Å and C6-C6'= 1.92 Å), but in this case both C5-C5' and C6-C6' distances immediately increase along the simulation in S_1 and a relaxation towards a point of intersection with S_0 leading to the dimer cannot be observed within the QM/MM time scale.

The fact that we cannot find an essentially barrierless path starting from an excited conformation, which corresponds to an accessible ground-state conformation, and leading to the dimer via a conical intersection, is due to the very high dimensionality of the phase space. Since the dimer is fully formed $\approx$ 1 ps after UV excitation [32], the limited time scale of QM/MM MD simulations is in principle not an obstacle to the observation of the dimerization. Though the use of the metadynamics approach increases the efficiency of the phase space sampling, the probability of finding such a path is still very low.

At the time of writing, a theoretical study by Robb et al. on thymine dimerization in the gas phase was published [161]. Their results show that the photoreaction occurs via a barrierless concerted mechanism on a singlet excited state. The latter mechanism takes place through a S_0/S_1 conical intersection, which is the funnel for ultrafast nonradiative decay leading to the thymine dimer. The structure of the S_0/S_1 conical intersection is characterized by the following distances: C5-C5'= 2.27 Å and C6-C6'= 2.17 Å. Robb et al. speculated that photoexcitation within DNA would lead to a spontaneous concerted [2 + 2] cycloaddition if the two neighboring thymines are at a configuration near the S_0/S_1 conical intersection geometry and suggested that such configurations are infrequent in the unexcited DNA, rationalizing the low dimerization quantum yield reported in a previous experiment [32]. Indeed, according to Figure 6.2, their S_0/S_1 conical intersection structure does not correspond to an accessible ground-state conformation.

6.4 Conclusion

In this chapter, we have presented a computational investigation of the thymine dimer formation in a DNA decamer on the singlet and triplet excited-states.

QM/MM metadynamics simulations in the triplet excited-state have provided some insight into the triplet mechanism of the thymine dimer formation, confirming the recently proposed stepwise pathway [85]. A reaction intermediate IT_T could be identified along the metadynamics simulation after crossing a barrier of around 17 kcal mol^{-1} from the basin of non-covalently linked thymine bases. This diradical intermediate is a precursor of the thymine dimer. Indeed, once the system is put back to the ground-state, we observe a barrierless formation of the dimer. Starting from a diradical structure which lies in the vicinity of IT_T but is not a precursor of the dimer, we could find a route on the triplet FES leading

6.4 Conclusion

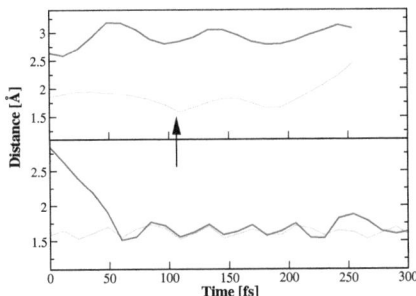

Figure 6.11: Time evolution of the C5-C5' bond length (blue) and the C6-C6' bond length (green) starting from IT_{S4} in S_1 (upper panel). The conformation with the shortest C6-C6' bond length (indicated by an arrow) along the simulation in S_1 is relaxed in S_0 (lower pannel), resulting in the thymine dimer formation.

to a crossing point with S_0 from where the system evolves towards the thymine dimer. The T_1/S_0 crossing points that we could identify in this study involve a biradical intermediate. However, the surface crossing region is not thermally accessible on a picosecond time scale from the well of non-covalently linked thymine bases in T_1 and does not correspond to any accessible ground-state conformation. A still unresolved issue is the possibility for the initially excited singlet state to undergo intersystem crossing to reach a triplet reactive intermediate. In a recent flash photolysis experiment, ISC to a triplet state in polymeric DNA could not be detected [143].

In the case of the singlet excited-state, a few species decaying directly to the thymine dimer once relaxing in S_0 could be identified. However, these species are out of reach since a high barrier (≈ 60 kcal/mol) must be overcome from the basin of non-covalently bound thymines in S_1 and no probable ground-state conformations can access them through photoexcitation. Nonetheless, some insight into the reaction mechanism could be gained by selecting a structure, IT_{S4}, in the vicinity of a precursor of the dimer. Starting from IT_{S4}, a route towards a crossing point with S_0 leading to the thymine dimer could be found. In contrast to the triplet mechanism, this pathway does not involve any stable intermediate. Thus, the reaction occurs in a concerted fashion and this rapid radiationless decay mechanism provides a rationale for the observation of the thymine dimer formation on a femtosecond time scale [32].

Ground-state simulations starting from S_1 and T_1 structures belonging to the basin of non-covalently linked thymine bases did not result in a decrease of both C5-C5' and C6-C6' distances and no thymine dimer formation could be observed. A complete pathway from an accessible excited conformation towards the thymine dimer via a crossing point with S_0 could not be uncovered since it would require much longer sampling time to span the full configurational space of this high dimensional system.

Acknowledgment

Computer resources were provided by the University of Zurich on the Matterhorn Cluster and by the Swiss National Supercomputing Center (CSCS) on IBM sp5.

Outlook

In this thesis the repair and formation reactions of the thymine dimer have been computationally investigated with the aim of providing a mechanistic picture of these two processes. Concerning the repair reaction, two distinct processes have been analyzed: the self-repair of the thymine dimer in DNA (Chapter 4) and the repair of the dimer in the active site of DNA photolyase (Chapter 5).

A quantum mechanics/molecular mechanics (QM/MM) approach to treat these problems is the method of choice since the environmental effects, which are crucial to obtain a correct representation of these reactions, are classically incorporated into the model. For example, hydrogen bonding from the complementary bases or the active-site residues to the dimer plays a crucial role in stabilizing the thymine dimer anion. The strain in the cyclobutane ring induced by the DNA or DNA photolyase is also important since it accelerates the splitting of the cyclobutane ring.

QM/MM simulations of the repair reaction revealed an asynchronously concerted mechanism for splitting of the ring, where C5-C5' bond cleavage spontaneously occurs after electron uptake and is subsequently followed by the C6-C6' bond breaking. Using metadynamics, an upper bound of 2.5 kcal/mol to the free energy barrier for the splitting of this bond is estimated. Our simulations provide for the first time a clear description of the splitting mechanism in the natural environments of the dimer. They confirmed both the thermodynamical and kinetic feasibility of the autocatalytic repair and the repair by DNA photolyase.

In the case of the enzymatic repair, key features characterizing the splitting mechanism have been disclosed by our simulations: continuous solvation reordering of the active site, a proton transfer from Glu283 to the thymine dimer and tight interactions between cationic side chains of Arg232 and Arg350 and the dimer. Thus, our study provides a better understanding of the catalytic mechanism that can hopefully lead to the design of improved artificial photolyases [162]. The design of new synthetic photolyases offers the long-term prospect of artificial DNA repair, which gained additional support by the recent demonstration of xenobiotic repair of thymine dimers in humans [124].

In Chapter 6, QM/MM metadynamics simulations in the excited states have provided some insight into the mechanism of thymine dimer formation. Our simulations showed that the triplet pathway towards the dimer involves a biradical intermediate, whereas the singlet mechanism does not proceed via any intermediate. Precursors of the thymine dimer could be identified on both surfaces, but a complete pathway from an accessible excited-state conformation towards the dimer via a crossing point with S_0 could not be uncovered. Further simulations could contribute to a full characterization of ground-state reactive conformations and to knowledge about the conformational criteria that make dimerization inevitable.

There are still many open questions concerning the mode of action of DNA photolyase. One example of an unresolved issue is whether the electron transfer pathway from the flavin cofactor to the lesion proceeds directly from the electron-donating π-system of the isoalloxazine ring or indirectly via the adenine moiety. Calculations of electron transfer rates is becoming possible within DFT due to recent advances [163, 164, 165, 166, 167, 168] and such theoretical studies could certainly contribute to clarify which of the two pathways is operational.

Acknowledgments

I am very grateful to Prof. Jürg Hutter for offering me a Ph.D. position in his group, for trusting me all the way through this long lasting effort, for his kindness and his generosity. I am indebted to Prof. Ursula Röthlisberger for defining the subject of my thesis, which allowed me to study a biological system, and agreeing on supervising it together with Jürg. I would like to thank Prof. Stefan Seeger and Prof. Peter Hamm for agreeing on being co-examiners of this thesis.

I'd like to thank all my present and former colleagues for the pleasant working atmosphere. First of all, my warmest thank goes to Teodoro Laino for helping me with crucial aspects in this work, for providing me guidance on the subtle art of writing scientific communications and his contagious enthusiasm for research. I'd like to thank Joost VandeVondele and Petra Munih for insightful discussions and kindly agreeing on proofreading large parts of this work. This work would have been definitively different without the cheerful presence of my office mate, Valéry Weber, who helped me in many ways and turned out to be the greatest "rock-punk-electro" DJ of the university! Special thanks to Thomas Kastl for his technical support and for sharing with me the joys and the pains of scientific research. The presence of Manuel Guidon, Samuele Giani and Florian Schiffmann was always a real pleasure and their help and good tips were always welcome. Ueli Feusi was also very helpful with many computer issues. I keep a nice memory as well of Marcella Iannuzzi, Thomas Chassaing, Teodora Todorova, Ari Seitsonen, Roger Nadler, Silvan Camenisch and Barbara Kirchner.

During the many opportunities to visit Ursula's group, I learnt to know many colleagues. I very much appreciated debatting scientific issues and spending great times at the summer- and ski outings with them. Special thanks go to Ivano for fruitful discussions. I'd like to thank from the bottom of my heart my friends Denis Bucher and Pascal Baillod for the unforgettable moments spent together and for their moral support throughout this research. I also had great relaxing times with Enrico Tapavicza, alias "Tierchen", either climbing in Spain or improvising a picnic in Vidy.

Very special thanks go to Axel Kohlmeyer who spent his time so generously trying to solve many of my QM/MM set-up problems. Despite corresponding via emails only, a true friendship developed between us and Axel offered me the opportunity to present my results in Philadelphia.

I am very grateful to my friends Iris, Nathalie and Catherine for their tender presence beside me. A special thought to my friends at Valdesia, my student society, who agreed on shifting many rehearsals of "Le Prologue" (the greatest play in Lausanne!) since I was coming from Zürich ! During these last four years, I also enjoyed a lot of great ski tours in Valais with the following crazy guys: Gilles, Caro, Zaz, Fonfon, Vinz and Big Fred.

Most importantly, I'd like to thank my parents, Henri and Hélène, and my brother, Pierre, for their constant support during all these years. I am very grateful to the family of Olivier, Henri, Françoise and Vanessa, for their continuous encouragements. The last thanks are for Olivier whose presence and love were of invaluable help to accomplish this work.

Curriculum Vitae

Name	Fanny Masson
Date of Birth	July 8, 1979
Place of Birth	Lausanne, Switzerland
Citizenship	Swiss

Education

October 2003 - present   Ph.D studies supervised by Prof. Dr. J. Hutter,
Physical Chemistry Institute,
University of Zürich,
and by Prof. Dr. Ursula Röthlisberger,
Laboratory of Computational Chemistry and Biochemistry,
Swiss Federal Institute of Technology EPFL, Lausanne.

March 2003   Master of Science Degree in Chemistry, with thesis in biochemistry,
Swiss Federal Institute of Technology EPFL, Lausanne.

2000-2001   Exchange year at the University of Ulm, Germany.

1998   Maturité Fédérale, section latin-science,
Gymnase Auguste Piccard, Lausanne.

Bibliography

[1] A. A. Vink and L. Roza. *Photochem. Photobiol. B: Biol.*, 65:101, 2001.

[2] H. Lodish, A. Berk, P. Matsudaira, C. A. Kaiser, M. Krieger, M. P. Scott, S. L. Zipursky, and J.Darnell. *Molecular Biology of the Cell*. WH Freeman: New York, 2004.

[3] A. Sancar. *Chem. Rev.*, 103:2203, 2003.

[4] R. Beukers and W. Berends. *Biochim. Biophys. Acta*, 41:550, 1960.

[5] A. Wacker, H. Dellweg, L. Träger, A. Kornhauser, E. Lodemann, G. Türch, R. Selzer, P. Chandra, and M. Ishimoto. *Photochem. Photobiol.*, 3:369, 1964.

[6] G. J. Fisher and H. E. Johns. *Photochemistry and Photobiology of Nucleic Acids*, pages 225–294. Academic Press, New York, 1976.

[7] A. M. Pedrini, S. Tornaletti, P. Menichini, and A. Abbondandolo. *Basic Life Sci.*, 38:295, 1986.

[8] D. A. Pearlman, S. R. Holbrook, D. H. Pirkle, and S.-H. Kim. *Science*, 227:1304, 1985.

[9] S. N. Rao, J. W. Keepers, and P. Kollman. *Nucleic Acids Res.*, 12:4789, 1984.

[10] K. Miaskiewicz, J. Miller, M. Cooney, and R. Osman. *J. Am. Chem. Soc.*, 118:9156, 1996.

[11] H. Liu, S. R. Hewitt, and J. B. Hays. *Genetics*, 154:503, 2000.

[12] H. Park, K. Zhang, Y. Ren, S. Nadji, N. Sinha, J.-S. Taylor, and C. Kang. *Proc. Natl Acad. Sci. USA*, 99:15965, 2002.

[13] A. Kelner. *Proc. Natl Acad. Sci. USA*, 35:73, 1949.

[14] R. Dulbecco. *Nature*, 163:949, 1949.

[15] C. S. Rupert, S. H. Goodgal, and R. M. Herriott. *J. Gen. Physiol.*, 41:451, 1958.

[16] A. Sancar and G. B. Sancar. *J. Mol. Biol.*, 172:223, 1984.

[17] G. B. Sancar, M. S. Jorns, G. Payne, D. J. Fluke, C. S. Rupert, and A. Sancar. *J. Biol. Chem.*, 262:492, 1987.

[18] Y.-T. Kao, C. Saxena, L. Wang, A. Sancar, and D. Zhong. *Proc. Natl Acad. Sci. USA*, 102:16128, 2005.

[19] J. Hahn, M.-E. Michel-Beyerle, and N. Rösch. *J. Phys. Chem. B*, 103:2001, 1999.

[20] D. B. Sanders and O. Wiest. *J. Am. Chem. Soc.*, 121:5127, 1999.

[21] J. Antony, D. M. Medvedev, and A. A. Stuchebrukhov. *J. Am. Chem. Soc.*, 122:1057, 2000.

[22] A. Mees, T. Klar, P. Gnau, U. Hennecke, A. P. M. Eker, T. Carell, and L.-O. Essen. *Science*, 306:1789, 2004.

[23] D. J.-F. Chinnapen and D. Sen. *Proc. Natl Acad. Sci. USA*, 101:65, 2004.

[24] M. R. Holman, T. Ito, and S.E. Rokita. *J. Am. Chem. Soc.*, 129:6, 2007.

[25] E. Sztumpf-Kulikowska, D. Shugar, and J. W. Boag. *Photochem. Photobiol.*, 6:41, 1967.

[26] D. W. Whillans and H. E. Johns. *Photochem. Photobiol.*, 9:323, 1969.

[27] A. A. Lamola and J. P. Mittal. *Science*, 154:1560, 1966.

[28] I. H. Brown and H. E. Johns. *Photochem. Photobiol.*, 8:273, 1968.

BIBLIOGRAPHY

[29] G. J. Fisher and H. E. Johns. *Photochem. Photobiol.*, 11:429, 1970.

[30] S. Y. Wang. *Fed. Proc. Fed. Am. Soc. Exp. Biol.*, 24:S71, 1965.

[31] J. Eisinger and R. G. Shulman. *Proc. Natl Acad. Sci. USA*, 58:895, 1967.

[32] W. J. Schreier, T. E. Schrader, F. O. Koller, P. Gilch, C. E. Crespo-Hernandez, V. N. Swaminathan, T. Carell, W. Zinth, and B. Kohler. *Science*, 315:625, 2007.

[33] E. Fermi, J. G. Pasta, and S. M. Ulam. Los Alamos LASL Report, 1995.

[34] B. Alder and T. Wainwright. *J. Chem. Phys.*, 27:1208, 1957.

[35] W. C. Swope, H. C. Andersen, P. H. Berens, and K. R. Wilson. *J. Chem. Phys.*, 76:637, 1982.

[36] D. Frenkel and B. Smit. *Understanding Molecular Simulation*. Academic Press, San Diego, 2nd edition, 2002.

[37] J. M. Wang, P. Cieplak, and P. A. Kollman. *J. Comp. Chem.*, 21:1049, 2000.

[38] T. A. Darden, D. York, and L. G. Pedersen. *J. Chem. Phys.*, 98:10089, 1993.

[39] U. Essmann, L. Perera, M. L. Berkowitz, T. Darden, H. Lee, and L. G. Pedersen. *J. Chem. Phys.*, 103:8577, 1995.

[40] J. W. Eastwood and R. W. Hockney. *J. Comp. Phys.*, 16:342, 1974.

[41] D. Marx and J. Hutter. *Ab initio Molecular Dynamics: Theory and Implementation, in Modern Methods and Algorithms of Quantum Chemistry*, volume 1. Forschungszentrum Jülich, 2000.

[42] G. Lippert, J. Hutter, and M. Parrinello. *Theor. Chem. Acc.*, 103:124, 1999.

[43] J. VandeVondele, M. Krack, F. Mohamed, M. Parrinello, T. Chassaing, and J. Hutter. *Comp. Phys. Comm.*, 167:103, 2005.

[44] The CP2K developers group. Freely available at the URL:: http://cp2k.berlios.de, released under GPL license., 2007.

[45] P. Hohenberg and W. Kohn. *Phys. Rev. B*, 136:864, 1964.

[46] W. Kohn and L. J. Sham. *Phys. Rev. A*, 140:1133, 1965.

[47] A. D. Becke. *Phys. Rev. A*, 38:3098, 1988.

[48] C. T. Lee, W. T. Yang, and R. G. Parr. *Phys. Rev. B*, 37:785, 1988.

[49] J. P. Perdew, K. Burke, and M. Ernzerhof. *Phys. Rev. Lett.*, 77:3865, 1996.

[50] A. D. Becke. *J. Chem. Phys.*, 98:5648, 1993.

[51] R. Car and M. Parrinello. *Phys. Rev. Lett.*, 55:2471, 1985.

[52] J. VandeVondele and J. Hutter. *J. Chem. Phys.*, 118:4365, 2003.

[53] A. Warshel and M. Levitt. *J. Mol. Biol.*, 103:227, 1976.

[54] O. A. von Lilienfeld, I. Tavernelli, U. Rothlisberger, and D. Sebastiani. *Phys. Rev. Lett.*, 93:153004, 2004.

[55] A. Laio, J. VandeVondele, and U. Rothlisberger. *J. Chem. Phys.*, 116:6941, 2002.

[56] T. Laino, F. Mohamed, A. Laio, and M. Parrinello. *J. Chem. Theory Comput.*, 2:1370, 2006.

[57] A. Laio and M. Parrinello. *Proc. Natl Acad. Sci. USA*, 99:12562, 2002.

[58] M. Iannuzzi, A. Laio, and M. Parrinello. *Phys. Rev. Lett.*, 90:238302, 2003.

[59] R. Martonak, A. Laio, and M. Parrinello. *Phys. Rev. Lett.*, 90:075503, 2003.

[60] A. Stirling, M. Iannuzzi, M. Parrinello, F. Molnar, V. Bernhart, and G. A. Luinstra. *Organometallics*, 24:2533, 2005.

[61] A. R. Oganov, R. Martonak, A. Laio, P. Raiteri, and M. Parrinello. *Nature*, 438:1142, 2005.

[62] A. Rodriguez-Fortea, M. Iannuzzi, and M. Parrinello. *J. Phys. Chem. B*, 110:3477, 2006.

[63] A. Laio, A. Rodriguez-Fortea, F. L. Gervasio, M. Ceccarelli, and M. Parrinello. *J. Phys. Chem. B*, 109:6714, 2005.

[64] I. Frank, J. Hutter, D. Marx, and M. Parrinello. *J. Chem. Phys.*, 108:4060, 1998.

[65] T. Ziegler, A. Rauk, and E. J. Baerends. *Theoret. Chim. Acta*, 43:261, 1977.

[66] S. Goedecker and C. J. Umrigar. *Phys. Rev. A*, 55:1765, 1997.

[67] E. Runge and E. K. U. Gross. *Phys. Rev. Lett.*, 52:997, 1984.

[68] E. K. U. Gross, J. Dobson, and M. Petersilka. *Topics in Current Chemistry: Density Functional Theory*. Springer, Berlin, 1996.

[69] R. B. Woodward and R. Hoffmann. *J. Am. Chem. Soc.*, 87:7556, 1965.

[70] P. F. Heelis, R. F. Hartman, and S. D. Rose. *Chem. Soc. Rev.*, 24:289, 1995.

[71] A. A. Voityuk, M.-E. Michel-Beyerle, and N. Rösch. *J. Am. Chem. Soc.*, 118:9750, 1996.

[72] A. A. Voityuk and N. Rösch. *J. Phys. Chem. A*, 101:8335, 1997.

[73] N. J. Saettel and O. Wiest. *J. Am. Chem. Soc.*, 123:2693, 2001.

[74] B. Durbeej and L. A. Eriksson. *J. Am. Chem. Soc.*, 122:10126, 2000.

[75] O. Dolgounitcheva, V. G. Zakrzewski, and J. V. Ortiz. *J. Phys. Chem. A*, 103:7912, 1999.

[76] I. Al-Jihad, J. Smets, and L. Adamowicz. *J. Phys. Chem. A*, 104:2994, 2000.

[77] *CPMD*, Version 3.11; The CPMD consortium, MPI für Festkörperforschung, and the IBM Zurich Research Laboratory. (see http://www.cpmd.org), 2006.

[78] M. Troullier and J. L. Martins. *Phys. Rev. B*, 43:1993, 1991.

[79] G. Martyna and M. Tuckerman. *J. Chem. Phys.*, 110:2810, 1999.

[80] S. Kirkpatrick, C. D. Gelatt, and M.P. Vecchi. *Science*, 220:671, 1983.

[81] S. Hirata and M. Head-Gordon. *Chem. Phys. Lett.*, 314:291, 1999.

[82] C. Dellago, P. G. Bolhuis, F. S. Csajka, and D. Chandler. *J. Chem. Phys.*, 108:1964, 1998.

[83] B. Durbeej and L. A. Eriksson. *Photochem. Photobiol. A: Chem.*, 152:95, 2002.

[84] W. Koch and M. C. Holthausen. *A Chemist's Guide to Density Functional Theory*. Wiley-VCH, 2000.

[85] R. B. Zhang and L. A. Eriksson. *J. Phys. Chem. B*, 110:7556, 2006.

[86] D. Grimberg, G. Segal, and A. Devaquet. *J. Am. Chem. Soc.*, 97:6629, 1975.

[87] J. Lorentzon, M. P. Fülscher, and B. O. Roos. *J. Am. Chem. Soc.*, 117:9265, 1995.

[88] M. K. Shukla and P. C. Mishra. *J. Chem. Phys.*, 240:319, 1999.

[89] C. E. Crespo-Hernandez, B. Cohen, P. M. Hare, and B. Kohler. *Chem. Rev.*, 104:1977, 2004.

[90] J. Cadet and P. Vigny. *in Bioorganic Photochemistry*, pages 1–272. Wiley, New York, 1990.

[91] K. M. Lima-Bessa and C. F. M. Menck. *Curr. Biol.*, 15:R58, 2005.

[92] L. M. Kundu, U. Linne, M. Marahiel, and T. Carell. *Chemistry*, 10:5697, 2004.

[93] V. I. Lyamichev, M. D. Frank-Kamenetskii, and V. N. Soyfer. *Nature*, 344:568, 1990.

[94] U. Schieferstein and F. Thoma. *Biochem.*, 35:7705, 1996.

[95] C. S. Rupert, S. H. Goodgal, and R. M. Herriott. *Journal of General Physiology*, 41:451, 1958.

[96] L. O. Essen and T. Klar. *Cell. Mol. Life Sci.*, 63:1266, 2006.

[97] B. Durbeej and L. A. Eriksson. *Photochem. Photobiol.*, 78:159, 2003.

[98] C. B. Harrison, L. L. O'Neil, and O. Wiest. *J. Phys. Chem. A*, 109:7001, 2005.

[99] N. A. Richardson, S. S. Wesolowski, and H. F. Schaefer. *J. Am. Chem. Soc.*, 124:10163, 2002.

[100] Y. Mantz, F. L. Gervasio, T. Laino, and M. Parrinello. *Phys. Rev. Lett.*, 99:058104, 2007.

[101] W. L. Jorgensen, J. Chandrasekhar, J. D. Madura, R. W. Impey, and M. L. Klein. *J. Chem. Phys.*, 79:926, 1983.

[102] D. A. Case and al. *AMBER8*. University of California, San Francisco, 2004.

[103] T. I. Spector, T. E. Cheatham III, and P. A. Kollman. *J. Am. Chem. Soc.*, 119:7095, 1997.

[104] T. E. Cheatham III, J. L. Miller, T. Fox, T. A. Darden, and P. A. Kollman. *J. Am. Chem. Soc.*, 117:4193, 1995.

[105] H. J. C. Berendsen, J. P. M. Postma, W. F. van Gunsteren, A. DiNola, and J. R. Haak. *J. Chem. Phys.*, 81:3684, 1984.

[106] T. Laino, F. Mohamed, A. Laio, and M. Parrinello. *J. Chem. Theory Comput.*, 1:1176, 2005.

[107] A. Schaefer, C. Huber, and R. Ahlrichs. *J. Chem. Phys.*, 100:5829, 1994.

[108] S. Goedecker, M. Teter, and J. Hutter. *Phys. Rev. B*, 54:1703, 1996.

[109] C. Hartwigsen, S. Goedecker, and J. Hutter. *Phys. Rev. B*, 58:3641, 1998.

[110] O. A. von Lilienfeld, I. Tavernelli, U. Rothlisberger, and D. Sebastiani. *J. Chem. Phys.*, 122:14113, 2005.

[111] I-C. Lin, O. A. von Lilienfeld, M. D. Coutinho-Neto, I. Tavernelli, and U. Rothlisberger. *To be published*, 2007.

[112] P. E. Bloechl. *J. Chem. Phys.*, 103:7422, 1995.

[113] S. J. Nose. *J. Chem. Phys.*, 81:511, 1984.

[114] W. G. Hoover. *Phys. Rev. A*, 31:1695, 1985.

[115] A. Laio, J. VandeVondele, and U. Rothlisberger. *J. Phys. Chem. B*, 106:7300, 2002.

[116] J. Rak, A. A. Voityuk, M.-E. Michel-Beyerle, and N. Rösch. *J. Phys. Chem. A*, 103:3569, 1999.

[117] C. Behrens, M. K. Cichon, F. Grolle, U. Hennecke, and T. Carell. *Topics in Current Chemistry: Long-Range Charge Transfer in DNA I*, pages 187–204. Springer, Berlin, 2004.

[118] R. F. Hartman, J. R. Van Camp, and S. D. Rose. *J. Org. Chem.*, 52:2684, 1987.

[119] P. F. Heelis. *J. Mol. Model.*, 1:18, 1995.

[120] C. Micheletti, A. Laio, and M. Parrinello. *Phys. Rev. Lett.*, 92:170601, 2004.

[121] L. Maragliano, A. Fischer, E. Vanden-Eijnden, and G. Ciccotti. *J. Chem. Phys.*, 125:24106, 2006.

[122] D. Branduardi, F. L. Gervasio, and M. Parrinello. *J. Chem. Phys.*, 126:054103, 2007.

[123] M. J. Frisch and al. *Gaussian 03*. Gaussian, Inc., Wallingford CT, 2004.

[124] H. Stege, L. Roza, A. A. Vink, M. Grewe, T. Ruzicka, S. Grether-Beck, and J. Krutmann. *Proc. Natl Acad. Sci. USA*, 97:1790, 2000.

[125] F. Masson, T. Laino, I. Tavernelli, U. Rothlisberger, and J. Hutter. *Submitted to J. Am. Chem. Soc.*, 2007.

BIBLIOGRAPHY

[126] A. W. MacFarlane IV and R. J. Stanley. *Biochem.*, 40:15203, 2001.

[127] S. M. Kapetanaki, M. Ramsey, Y. M. Gindt, and J. P. M. Schelvis. *J. Am. Chem. Soc.*, 126:6214, 2004.

[128] I. Husain, G. B. Sancar, S. R. Holbrook, and A. Sancar. *J. Biol. Chem.*, 262:13188, 1987.

[129] B. J. Vande Berg and G. B. Sancar. *J. Biol. Chem.*, 273:20276, 1998.

[130] S. R. Yeh and D. E. Falvey. *J. Am. Chem. Soc.*, 113:8557, 1991.

[131] S. R. Yeh and D. E. Falvey. *J. Am. Chem. Soc.*, 114:7313, 1992.

[132] C. Pak, Kubo J., Majima T., and H. Sakurai. *Photochem. Photobiol.*, 36:273, 1982.

[133] R. Epple, E.-U. Wallenborn, and T. Carell. *J. Am. Chem. Soc.*, 119:7440, 1997.

[134] A. Noy, A. Perez, F. Lankas, F. J. Luque, and M. Orozco. *J. Mol. Biol.*, 343:627, 2004.

[135] W. Domcke. *Femtochemistry*, pages 133–145. Wiley-VCH, Weinheim, 2001.

[136] F. Bernardi, S. De, M. Olivucci, and M. A. Robb. *J. Am. Chem. Soc.*, 112:1737, 1990.

[137] F. Bernardi, M. Olivucci, and M. A. Robb. *Pure & Appl. Chem*, 67:17, 1995.

[138] I. G. Gut, P. D. Wood, and R. W. Redmond. *J. Am. Chem. Soc.*, 118:2366, 1996.

[139] P. D. Wood and R. W. Redmond. *J. Am. Chem. Soc.*, 118:4256, 1996.

[140] M. T. Nguyen, R. Zhang, P.-C. Nam, and A. Ceulemans. *J. Phys. Chem. A*, 108:6554, 2004.

[141] C. Salet, R. Bensasson, and R. S. Becker. *Photochem. Photobiol.*, 30:325, 1979.

[142] Q. Song, W. Lin, S. Yao, and N. Lin. *Photochem. Photobiol. A: Chem.*, 114:181, 1998.

[143] S. Marguet and D. Markovitsi. *J. Am. Chem. Soc.*, 127:5780, 2005.

[144] V. Lhiaubet-Vallet, M. C. Cuquerella, J. V. Castell, F. Bosca, and M. A. Miranda. *J. Phys. Chem. B*, 111:7409, 2007.

[145] S. Grimm, C. Nonnenberg, and I. Frank. *J. Chem. Phys.*, 119:11574, 2003.

[146] C. Nonnenberg, S. Grimm, and I. Frank. *J. Chem. Phys.*, 119:11585, 2003.

[147] C. Nonnenberg, C. Brauchle, and I. Frank. *J. Chem. Phys.*, 122:014311, 2005.

[148] H. Langer and N. L. Doltsinis. *J. Chem. Phys.*, 118:5400, 2003.

[149] U. F. Röhrig, I. Frank, J. Hutter, A. Laio, J. VandeVondele, and U. Rothlisberger. *ChemPhysChem*, 4:1177, 2003.

[150] H. Langer, N. L. Doltsinis, and D. Marx. *ChemPhysChem*, 6:1734, 2005.

[151] P. R. L. Markwick, N. L. Doltsinis, and J. Schlitter. *J. Chem. Phys.*, 126:045104, 2007.

[152] W. R. P. Scott, P. H. Hünenberger, I. G. Tironi, A. E. Mark, S. R. Billeter, J. Fennen, A. E. Torda, T. Huber, P. Krüger, and W. F. van Gunsteren. *J. Phys. Chem. A*, 103:3596, 1999.

[153] P. H. Hünenberger. *J. Chem. Phys.*, 113:10464, 2000.

[154] M. Odelius, D. Laikov, and J. Hutter. *J. Mol. Struc. (Theochem)*, 630:163, 2003.

[155] J. Hutter. *J. Chem. Phys.*, 118:3928, 2003.

[156] D. Varsano, R. DiFelice, M. A. L. Marques, and A. Rubio. *J. Phys. Chem. B*, 110:7129, 2006.

[157] O. Treutler and R. Ahlrichs. *J. Chem. Phys.*, 102:346, 1995.

[158] R. Bauernschmitt and R. Ahlrichs. *Chem. Phys. Lett.*, 256:454, 1996.

BIBLIOGRAPHY

[159] R. Bauernschmitt and R. Ahlrichs. *J. Chem. Phys.*, 104:9047, 1996.

[160] C. M. Marian, F. Schneider, M. Kleinschmidt, and J. Tatchen. *Eur. Phys. J. D*, 20:357, 2002.

[161] M. Boggio-Pasqua, G. Groenhof, L. V. Schäfer, H. Grubmüller, and M. A. Robb. *J. Am. Chem. Soc.*, 129:10996, 2007.

[162] O. Wiest, C. B. Harrison, N. J. Saettel, R. Cibulka, M. Sax, and B. König. *J. Org. Chem.*, 69:8183, 2004.

[163] Wu Q. and Van Voorhis T. *J. Chem. Phys.*, 125:164105, 2006.

[164] M. Cascella, A. Magistrato, I. Tavernelli, P. Carloni, and U. Rothlisberger. *Proc. Natl Acad. Sci. USA*, 103:19641, 2006.

[165] VandeVondele J., Lynden-Bell R., Meijer E. J., and M. Sprik. *J. Phys. Chem. B*, 110:3614, 2006.

[166] VandeVondele J., Sulpizi M., and Sprik M. *Angew. Chem. Int. Ed.*, 58:895, 2006.

[167] VandeVondele J., Sulpizi M., and Sprik M. *CHIMIA*, 61:155, 2007.

[168] Sulpizi M., Raugei S., VandeVondele J., Carloni P., and Sprik M. *J. Phys. Chem. B*, 111:3969, 2007.

VDM Verlagsservicegesellschaft mbH

Die VDM Verlagsservicegesellschaft sucht für wissenschaftliche Verlage abgeschlossene und herausragende

Dissertationen, Habilitationen, Diplomarbeiten, Master Theses, Magisterarbeiten usw.

für die kostenlose Publikation als Fachbuch.

Sie verfügen über eine Arbeit, die hohen inhaltlichen und formalen Ansprüchen genügt, und haben Interesse an einer honorarvergüteten Publikation?

Dann senden Sie bitte erste Informationen über sich und Ihre Arbeit per Email an *info@vdm-vsg.de*.

Sie erhalten kurzfristig unser Feedback!

VDM Verlagsservicegesellschaft mbH
Dudweiler Landstr. 99
D - 66123 Saarbrücken
www.vdm-vsg.de

Telefon +49 681 3720 174
Fax   +49 681 3720 1749

Die VDM Verlagsservicegesellschaft mbH vertritt

Printed by Books on Demand GmbH, Norderstedt / Germany